P9-BZP-153

Rum Maniacs

Rum Maniacs

Alcoholic Insanity in the Early American Republic

MATTHEW
WARNER OSBORN

The University of Chicago Press
Chicago and London

Matthew Warner Osborn is assistant professor of history at the University of Missouri–Kansas City.

The University of Chicago Press, Chicago 60637
The University of Chicago Press, Ltd., London

Printed in the United States of America

23 22 21 20 19 18 17 16 15 14   1 2 3 4 5

ISBN-13: 978-0-226-09989-7 (cloth)
ISBN-13: 978-0-226-09992-7 (e-book)

DOI: 10.7208/chicago/9780226099927.001.0001

Library of Congress Cataloging-in-Publication Data

Osborn, Matthew Warner, author.
Rum maniacs : alcoholic insanity in the early American Republic / Matthew Warner Osborn.
pages ; cm
Includes bibliographical references and index.
ISBN 978-0-226-09989-7 (cloth : alk. paper) — ISBN 978-0-226-09992-7 (e-book) 1. Delirium tremens—United States—History—19th century. 2. Delirium tremens—Pennsylvania—Philadelphia—History—19th century. 3. Delirium tremens—United States—Psychological aspects—History—19th century. 4. Delirium tremens—Social aspects—United States—History—19th century. 5. Temperance—United States—History—19th century. I. Title.
RC526.O83 2014
362.292—dc23

2013025544

♾ This paper meets the requirements of ANSI/NISO Z39.48-1992 (Permanence of Paper).

CONTENTS

ACKNOWLEDGMENTS

This book benefited from the intellectual, financial, and emotional generosity of many people. Karen Halttunen's scholarship was an important inspiration. From beginning to end, her mentoring and friendship proved invaluable. Alan Taylor, David Henkin, Catherine Kudlick, and Daniel Richter read drafts of the entire manuscript. Colleagues and friends took time to offer suggestions, read chapters, and bear with my handwringing, especially Aaron Wunsch, Amanda Moniz, Brad Cazden, Jessica Roney, Joe Karten, Julie Kim, Justine Murrison, Robert Weiss, Rob Habberman, Tim Yates, and Trisha Posey. Two anonymous readers for the University of Chicago Press provided me with detailed and useful comments. Alice Bennett smoothed many rough edges. Robert Devens offered patient encouragement and sage advice through my prolonged revisions.

I have presented material from the manuscript in various formal and informal venues and learned much from audience members. Bruce Dorsey, Kyle Bulthuis, Kyle Volk, Pat Cohen, Renate Wilson, Ric Caric, Richard Bell, Scott Martin, and William J. Rorabaugh participated with me on academic panels that allowed me to test new ideas. For their invitations, I thank Michael Sappol and the National Library of Medicine, Daniel Richter and the McNeil Center for Early American Studies, the Clendening Library, and Philadelphia's Edgar Allan Poe House.

I received help from numerous archivists and librarians, especially Stacey Peeples at the Pennsylvania Hospital Historical Collections; John Pollack in the Rare Books and Manuscript Library at the University of Pennsylvania; Roy Goodman and J. J. Ahern at the American Philosophical Society; James Green, Cornelia King, and Wendy Woloson at the Library Company of Philadelphia; Edward Mormon, then director of the Historical Medical

Library at the College of Physicians of Philadelphia; and Carla Lillvik at Harvard University's Gutman Library.

Many people shared research and resources. Karla Kelling was enormously generous in helping me with research into the Philadelphia Almshouse. Sarah Knott and Ric Caric allowed me to cite their unpublished work. The Magic Lantern Society of the United States and Canada responded to a plea for help, and Terry Borton of the American Magic-Lantern Theater shared slides from his private collection. I hope the little work I have done here on the phantasmagoria will encourage scholarship on this fascinating theatrical form.

Over the course of my research and writing, I met a number of people who have suffered delirium tremens or witnessed the affliction. I want to thank the several who volunteered difficult memories for giving me some insight into the lived experience of the disease.

I began this project while on a one-year fellowship from the Humanities Institute at the University of California–Davis. A grant from the UCD history department's Roland Marchand Memorial Fund enabled me to write an essay that received the Roy Porter Student Essay Prize from the Society for the Social History of Medicine. That essay was subsequently published in the society's journal. The bulk of this research was funded by three grants. I received a Barra Foundation Fellowship from the McNeil Center for Early American Studies. The Library Company of Philadelphia awarded me a Roy M. Greenfield Fellowship, and I spent a wonderful year as a denizen of the Cassatt House. A research fellowship from the American Philosophical Society enabled me to enjoy a snowy February in its beautiful library. The UCD history department granted me two yearlong fellowships that enabled me to complete an earlier version. During that time, I adapted a portion of my research for publication in the *Journal of the Early Republic*. That material makes up the bulk of chapter 3.

While I was revising the manuscript, Occidental College provided me a safe harbor. Dolores Trevizo, Lisa Sousa, Lynn Dumenil, Sharla Fett, and everyone in the history department offered me friendship and support. I finished the book as a member of the history department at the University of Missouri–Kansas City.

My mother, Becky Osborn Coolidge, and stepfather, John S. Coolidge, always expressed a confidence that I often sorely needed. Erin Merritt read everything and offered limitless curiosity, belief, and intelligence. Our young children, Paul and Imogen, often asked when my book would be done. I am relieved to finally tell them that this particular one is finished.

INTRODUCTION

On a hot summer afternoon in 1849, Edgar Allan Poe appeared unexpectedly at the home of his friend John Sartain. Poe looked "pale and haggard," Sartain would later recount, with "a wild and frightened expression in his eyes." Convinced that murderers were pursuing him, Poe feverishly described being thrown into Philadelphia's Moyamensing Prison, where he experienced bizarre hallucinations. From his cell, he saw a woman standing on top of the prison's tower, "a young female brightly radiant, like silver dipped in light." She tried to entrap him with a series of questions, but Poe steadfastly refused to answer. Soon after, a prison guard led him to a boiling cauldron and insisted that he take a drink, but Poe resisted the guard's murderous intentions. The plot took a horrifying turn when Poe's tormentors dragged his mother before him and began to mutilate and dismember her. Describing to Sartain the gruesome vision of his mother's legs being sawed off, Poe collapsed into a convulsion.[1] After his recovery, Poe understood his hallucinations and delusions to have been symptoms of a disease caused by a bout of heavy drinking. He later referred to these experiences in a letter to a close family member. "For more than ten days I was totally deranged, although I was not drinking one drop," he wrote. "During this interval I imagined the most horrible calamities. . . . All was hallucination, arising from . . . an attack of *mania-a-potu*."[2]

Better known as delirium tremens, mania a potu was common in nineteenth-century hospitals. The disease was also a subject of romantic speculation. That Poe suffered from it was both tragic and eerily appropriate, given the nature of his writing. Commenting on Poe's experience, one anonymous commentator noted that in his agitation, "the poet seemed a personification of his own 'Raven.'"

> Caught from some unhappy master whom unmerciful Disaster
> Followed fast and followed faster till his songs one burden bore—
> Till the dirges of his Hope that melancholy burden bore
> Of "Never—nevermore."[3]

This comparison had first been made in 1848, when an editor criticized Poe's famous poem for "wild and unbridled extravagance" and wondered if the author had intended it as a description of the "fantastic terrors which afflict a sufferer from *delirium tremens*."[4]

Rum Maniacs traces how and why heavy drinking became a subject of medical interest, social controversy, and lurid fascination in the early American republic. At the heart of that story is the history of delirium tremens and the "fantastic terrors" that characterize it. Whether or not Poe intended *The Raven* to evoke the disease, in the mid-nineteenth century delirium tremens had inspired a wide range of popular theater, poetry, fiction, and illustration. It was a relatively new disease, however. British physicians had first described it just three decades earlier, in 1813. Doctors in the United States began studying the disease the following year.[5] This development marked the beginning of the dramatic intervention of the American medical profession into the social response to alcohol abuse, or "intemperance" as it was termed then. Delirium tremens changed how the medical profession observed, understood, and treated the more general problem of alcohol abuse. Indeed, the delirium tremens diagnosis became the foundation for the medical conviction and popular belief that habitual heavy drinking was pathological—a self-destructive compulsion that constituted a psychological and physiological disease.

Several months after his incarceration in Moyamensing Prison, and again suffering hallucinations, Poe died at Washington Medical College in Baltimore in a section of the hospital reserved for inebriates.[6] That he died in these circumstances was one historical consequence of the delirium tremens diagnosis. Physicians had long recognized that heavy drinking damaged health, but before 1813 they had little interest in treating those overcome by intoxication or suffering the violent symptoms of delirium tremens. In the years following the Revolution, the nation's most prominent physicians described intemperance as a dire threat to the nation's physical and moral health and a pressing danger to fragile republican institutions. Despite this concern, drunkards went largely untreated. City authorities confined them to dank cells at the almshouse or in jail, where they received little if any medical attention. By the 1820s, cases of delirium tremens began appearing regularly in medical journals, hospital records, and death statistics. Nationwide, inebriates were increasingly put in hospital beds, diagnosed, and treated. In countless postmortem examinations, physicians studied the morbid effects of heavy drinking on the internal organs. As the century progressed, newly built hospitals set aside whole wards for treating drunkards. For the physicians at Washington Medical College, inebriate patients suffering hallucinations and delusions would have been common.

Delirium tremens remains common in hospitals, though current medical definitions of the disease are narrower than in the early nineteenth century. The US National Library of Medicine and the National Institutes of Health describe delirium tremens as a disease that can follow sudden withdrawal of alcohol.[7] When heavy drinkers lessen their intake or stop drinking altogether, they are at risk for symptoms that include tremors, anxiety, nightmares, and vomiting. Delirium tremens is particularly characterized by "severe confusion and visual hallucinations."[8] In the early nineteenth century, physicians commonly identified the disease with mental aberrations and habitual heavy drinking, but not always with alcohol withdrawal. The debate over the relation between withdrawal and delirium tremens endured well into the twentieth century.[9] In part this debate arose because the connection is not altogether obvious. Heavy drinkers do not need to stop drinking to develop hallucinations and other withdrawal symptoms. Early nineteenth-century doctors understood delirium tremens to be insanity caused by habitual heavy drinking, and many also noted that it commonly occurred when a drunkard suddenly abstained. In the earliest published case histories, physicians identified the disease by its characteristically violent symptoms, which included trembling, vomiting, paranoia, and, especially, vivid hallucinations.

What does it mean to say that doctors "first described" delirium tremens in 1813? Why did this particular disease become a compelling subject of interest in the American medical community and in popular culture? Although the histories of drinking, intemperance, and alcohol addiction have attracted a wide range of scholarship, only a few historians have addressed these questions about delirium tremens.[10] Mid-twentieth-century historians of medicine explained the distinction of delirium tremens from other forms of insanity as the result of an increasing sophistication in charting the health consequences of alcohol abuse and classifying mental disorders.[11] According to this Whiggish argument, as physicians studied lunatics they simply became more adept at distinguishing individual mental disorders such as delirium tremens.[12] The discovery of the disease was thus the result of the natural development and expansion of medical knowledge.

This view leaves many difficult questions unanswered. In the 1810s, for instance, the pathological condition that delirium tremens described was certainly not a new discovery. Eighteenth-century doctors were well aware that habitual heavy drinking could lead to insanity. Physicians at the Philadelphia almshouse were quite familiar with the condition years before they adopted the new diagnosis. Further, evidence demonstrates that the delirium tremens diagnosis was not "progress," at least in terms of patient care. Its advent actually resulted in more damaging treatments for much of the nineteenth century. Why, then, did a well-known condition that had long held no interest for physicians suddenly become a cutting-edge medical diagnosis? Why would physicians adopt the diagnosis when it did not bring a cure for the disease and in fact led to treatments far more harmful to the patient? What exactly was "new" about delirium tremens?

The delirium tremens diagnosis describes a set of symptoms that derive from a biological mechanism—a perilous condition brought on by excessive drinking—but those symptoms are open to a range of interpretations and descriptions that are historically contingent.[13] Giving a new name to a well-known condition, the diagnosis remade it into a significant, even fascinating, disease. Delirium tremens was made possible in part by broad developments within the medical profession, including the transatlantic circulation of medical texts and journals, the rapid expansion of medical education, and the growing practice of pathological anatomy. But physicians' preoccupation with delirium tremens had much to do with historical developments that lay outside the medical sphere. The imagery in physicians' narrative case histories, for instance, and the intellectual categories they used to describe the disease derived from contemporary trends in literary and popular culture, including the spread of romantic theories of the mind, changing conceptions of deviance and radical evil, and literary and popular romanticism. Using the language of romanticism, physicians ascribed a profound social significance to the disease. Case histories and medical records link physicians' interest in delirium tremens to broader concerns with urban poverty, economic instability, and social fluidity. This new disease was thus inseparable from intellectual, social, economic, and cultural developments of the late eighteenth and early nineteenth centuries.

The history of delirium tremens illuminates how a form of human suffering became a compelling topic of medical interest, with far-reaching consequences for American medicine, society, and culture. One consequence was the medical and popular conviction that heavy drinking could itself be a disease. Based on their postmortem examinations of delirium tremens victims, physicians theorized that the inebriates' internal organs could become habituated to alcoholic stimulation. In publicizing their new discoveries, physicians cited their physiological findings as biological evidence for the common observation that drunkards had an overwhelming craving for drink. These findings were spread through popular health journals, magazines, public lectures, and temperance organizations, and pathological anatomy formed the foundation for the claim that heavy drinkers suffered from a physiological and psychological compulsion that could quickly throw the unsuspecting drinker into a state of insanity. Ultimately, forms of mass culture, especially fiction and theatrical entertainment, established delirium tremens in popular consciousness, shaping a new public awareness that the habit of heavy drinking could in fact be a physical affliction.

These were decades when "intemperance" became an enormously controversial social issue. Prominent citizens blamed drinking for a host of frightening problems, including the rapid growth of urban poverty, epidemic disease, and social disorder. Newspapers and magazines commonly related how alcohol drove individuals, especially young white men, into ill health, social disgrace, poverty, and moral depravity, and even to shocking evil: murder, torture, rape, and suicide. Scholars of the early American republic have long studied these controversies surrounding alcoholic drink to highlight the development of explosive political and social tensions.[14] They have focused especially on the membership, activism, and literature of temperance societies. In 1826 the national temperance movement blossomed after the founding of the American Temperance Society, which became a national hub for a growing number of local groups dedicated to publicizing the dangers of drink. Pointing to the involvement of wealthy entrepreneurs, Christian evangelicals, and socially ambitious men and women, historians have linked the popularity of temperance societies to the rise of wage labor and capitalist production, the spread of new forms of evangelical Christian devotion, and the transformation of the northern white middle-class family. Temperance, piety, and industry were modes of behavior that shaped the social distinctiveness of the new middle class taking shape in the 1820s and 1830s. For the wealthy entrepreneurs who provided much of the financial backing for anti-alcohol activism, temperance was also part of an effort to mold a more industrious male workforce to fit a new capitalist labor regime.[15]

Characterizing temperance as a "moral" reform movement, historians have tended to dismiss physicians' claims about the health consequences of heavy drinking as compelled by ideology. In this view, "temperance physicians" were simply "dressing up drink discourse ideas in scientific language," as one historian has put it.[16] But even the most influential physicians in the temperance movement were doctors before they were activists. The medical profession was heavily involved in temperance societies, and doctors' support reflected the particular professional imperatives they faced in the medical marketplace. These men of science aspired to be elite professionals, and their work with temperance organizations derived from their training and served their own ambitions. *Rum Maniacs* thus contributes to an understanding of how attitudes toward alcohol and intoxication expressed developments in American society and culture that extended beyond moral reform movements, Christian evangelicals, or struggles over new forms of labor discipline. Rather than simply responding to the imperatives of moral reform, physicians played a fundamental role in shaping popular concerns and conceptions of heavy drinking and its consequences.[17]

One of the central underlying historical questions raised by the enormous popularity of temperance societies is, How and why did certain groups in society redefine problem drinking? What inspired Americans to begin to describe certain consumption patterns as pathological?[18] Despite being skeptical of the sincerity of early American physicians, historians have long noted the powerful influence of medical ideas in shaping the ideology of temperance organizations. In *The Alcoholic Republic: An American Tradition* (1979), W. J. Rorabaugh argued that medical science during the eighteenth-century Scottish Enlightenment was the most important intellectual development underlying the temperance movement.[19] Most important, he argued, the Philadelphia physician Benjamin Rush's essay *An Inquiry into the Effects of Spirituous Liquors on the Human Body* (first published in 1784), provided a new and convincing argument for ministers, moralists, politicians, and social reformers who advocated temperance. In his history of evangelical reformers, Robert Abzug has also noted that Rush's pamphlet and new early nineteenth-century medical research on physiology were central to the widespread support for the temperance cause.[20] Relatively little has been written on the historical relation between medicine and temperance in the early republic.[21]

Focusing closely on physicians' involvement in the social response to alcohol abuse, *Rum Maniacs* illuminates how the medical profession developed in the eclectic and competitive marketplace of the early republic. This book looks especially at Philadelphia, the unquestioned center of American medicine during these years. The University of Pennsylvania, far and away the country's largest and most prestigious medical school, was one of the city's two publicly chartered medical universities, five private medical institutes, two hospitals, and three public dispensaries where medical students could gain instruction. Each year, over five hundred young men flocked to Philadelphia for medical training, far more than to any other American city. In the Western world, only the great universities at Paris and Edinburgh attracted more students. The nation's first hospital, the Pennsylvania Hospital, was founded in Philadelphia, and it maintained the nation's first medical library, giving students and physicians access to the latest European medical literature. By the 1830s the Philadelphia Almshouse hospital had become the largest in the United States. Nationally, it was the most prestigious institution where medical students could gain clinical experience. Philadelphia also boasted the most influential, nationally circulating medical journals and easy access to cadavers, which provided students with the all-important experience of postmortem dissection. Before the Civil War, Philadelphia's medical professors wrote almost every major textbook used in American medical schools.[22]

Beginning in the 1770s and continuing through the nineteenth century, the growing American medical profession harbored an enduring concern with the significance and social consequences of heavy drinking. Philadelphia physicians played a leading role in shaping medical conceptions of pathological drinking both within the profession and among the general public. No individual was more significant than Benjamin Rush, the nation's most eminent physician. A signer of the Declaration of Independence, he was the most popular and influential professor of medicine at the University of Pennsylvania from 1790 until his death in 1813.[23] In the 1810s and 1820s, most of the growing literature on delirium tremens was written by Philadelphia physicians or published in the city's medical journals. In the national temperance movement as well, the physicians most active in speaking and writing for anti-alcohol organizations received their training at the University of Pennsylvania.[24] Philadelphia was a national center of activism, and elite physicians were leaders of the city's temperance organizations. In published temperance essays, popular health publications, and public lectures, these physicians popularized medical conceptions of alcohol abuse developed in Philadelphia's lecture halls and hospitals.

Philadelphia is also an ideal place to study how sweeping changes within the American medical profession interacted with socioeconomic tensions associated with capitalist transformation in the early republic. In addition to being the heart of American medicine, the city stood at the center of national political, intellectual, and cultural life. In the 1790s, Philadelphia was the nation's most populous city and its second largest port; it was home to an enormously wealthy merchant community and served as the nation's temporary capital. When the city's port declined after 1815, Philadelphia remade itself into a leading center of industry, second in size only to New York, and until the 1840s it was the nation's banking center.[25] As was typical of urban areas during this period, Philadelphia and its surrounding suburbs experienced massive population growth, from approximately 44,000 in 1790 to over 388,000 by 1850.[26] These socioeconomic transformations were marked by a series of crises—epidemics, financial panics, economic depressions, and the near collapse of poor relief services—that influenced the development of the medical profession. Medical responses to intemperance were shaped by the rapid growth of urban poverty, new social and economic imperatives that accompanied the boom-and-bust market economy, and new disparities of wealth and social status. The preoccupation with delirium tremens, as well as the more general health consequences of intemperance, in large part reflected the social aspirations of ambitious young medical men striving for social respectability and economic advance in these difficult and uncertain decades.

The chapters that follow are both thematic and loosely chronological. The first three focus on the Philadelphia medical community's relation to the intellectual, social, and cultural context of the early republic. Chapter 1 traces the intellectual course of Benjamin Rush's views on intemperance and on human physiology and pyschology. The chapter places the evolution of Rush's thinking about ardent spirits in the context of his political commitments, intellectual developments within the international medical community, and changes in popular culture. Delirium tremens, first described in America at the Philadelphia Almshouse in 1814, is the primary subject of chapter 2. The chapter addresses how the rapid growth of the medical profession, the influence of new European medical theories and practices, and the profound economic depression that followed the Panic of 1819 shaped physicians' interest in the new disease. Chapter 3 constructs a social history of inebriates by drawing on a sample of over 1,500 individuals who died of alcohol abuse in Philadelphia between 1825 and 1850. The chapter links medical concerns with pathological drinking to the rapid growth of urban poverty and widening class differences.

The final three chapters chart how physicians and their medical theories shaped cultural conceptions of temperate and pathological consumption. Chapter 4 documents the central role physicians played in the national temperance movement that blossomed in the late 1820s. It argues that physicians' temperance activism was central to an effort to remake the American medical profession in response to the imperatives of the competitive market. Temperance societies were vehicles through which physicians linked conventional medicine with health, social respectability, and economic well-being as they tried to make conventional medical theory relevant to a populace increasingly suspicious of elite professionals. Chapter 5 charts physicians' midcentury attempts to warn of the frightening consequences of intemperance, presenting habitual heavy drinking as an overwhelming physiological compulsion. It explains how physicians came to describe such drinking as a disease and why they had little inclination to develop therapies or treat those who suffered from it. The final chapter traces how delirium tremens shaped representations of pathological drinking in mainstream popular culture. Exploring the symbolic dimensions of the disease, the chapter describes how and why this ugly affliction became compelling theater for middle-class audiences.

One of the central problems this book addresses concerns language and seeing. Today a physician who sees a heavy drinker exhibiting paranoia, hallucinations, trembling limbs, and violent puking might describe that person as suffering from alcohol withdrawal or, if the symptoms are acute, even delirium tremens. In the eighteenth century, physicians might refer to such symptoms in very general terms, like mania, or simply note that the patient had become furious, without feeling it necessary to describe or classify the condition more specifically. Medical terms and concepts change over time, and they also take on specific meanings in particular historical contexts. When filling out death certificates, nineteenth-century physicians knew this very well. As I will discuss in chapter 3, "delirium tremens," "intemperance," and "brain fever" could all describe the same fatality, yet they carried very different social implications and consequences.

I have tried to define and use terms common in the nineteenth century, such as "delirium tremens," "intemperance," "inebriate," and "drunkard," as they were used during the period, but some remain in use today, though their meanings have changed. First used by the British physician Thomas Sutton in 1813, "delirium tremens" eventually became the dominant term for describing the disease.[27] Nineteenth-century physicians used an array of interchangeable terms to refer to this condition, such as mania a potu, mania a temulentia, and the "brain fever of drunkenness." At times I will draw on modern terminology to more clearly describe how medical terms, theories, and practices changed over time. I will use "alcoholic insanity," for instance, to refer to insanity caused by excessive drinking or sudden abstinence from long-term chronic drinking. By the 1830s, delirium tremens was the term physicians most often used, although mania a potu survived at least until the Civil War.

I use "alcohol addiction" to describe compulsive drinking, but for the most part I avoid the modern "alcoholism," a particular disease model developed by physicians in the late nineteenth and twentieth centuries. The word came into common use in the twentieth century.[28] Because the term is so bound up in contemporary popular conceptions of pathological drinking, I generally omit it to avoid coloring nineteenth-century notions of "intemperance" with our contemporary ideas about alcoholism. Instead, I will often use the more general terms "alcohol abuse" and "pathological drinking" loosely to refer to problematic, unhealthy, and damaging drinking patterns. In using these terms, I do not intend (nor does my evidence allow me) to make any claims about what constitutes unhealthy or problem drinking. Indeed, the historical construction of "pathological drinking"—what it looks like, its consequences, and its cultural significance—is a central topic of this book.

The history of delirium tremens in the early republic casts light on the social and cultural significance of medicine and disease during a period of rapid socioeconomic change. In the hothouse atmosphere of Philadelphia medical schools, delirium tremens became a subject of intense interest and profound meaning. In professional journals, anatomy theaters, and university lecture halls, elite physicians and their students painted vivid portraits of the disease's horrors that rivaled the supernatural imagery permeating the era's popular gothic novels and theater. Young doctors wrote case histories filled with detailed and fanciful descriptions of their patients' hallucinations: ghosts, devils, vermin, and other frightening visions. These men chose to highlight these hallucinations, delusions, and other violent physical symptoms because the disease had become a sort of metaphoric theater. In the midst of the most profound economic depression the nation had ever experienced, physicians and medical students associated delirium tremens with bankruptcy, business failure, and social downfall. Composed largely of young men seeking an uncertain stake in bourgeois society, this all-male cadre worked in Philadelphia institutions that catered primarily to the poor and indigent. The disease became especially meaningful to these aspiring doctors as they daily confronted the ravages of intemperance and economic failure. In delirium tremens, they described a condition that was as much a disease of social downfall as a deadly consequence of heavy drinking.

New medical beliefs and practices powerfully reinforced emerging social distinctions, especially along the lines of class and gender. Physicians' efforts reflected their own social backgrounds, as well as the values and worldview of the middle-class patients they hoped to win. Speaking in the disinterested language of science, in the universal terms of human health, physicians detailed the physiological basis of an emerging middle-class ethos. Drinking was not the only vice that American doctors denounced; they also targeted tobacco, opium, masturbation, and rich foods. Physicians testified to many health dangers, but alcohol was their most important concern. University-trained physicians asserted professional authority based on the social usefulness and individual utility of their orthodox medicine. Their intervention into the social response to alcohol abuse was at the heart of an effort to shape a new way forward for a young and rapidly growing American medical profession.

The social and cultural consequences of the impulse to pathologize heavy drinking were ambiguous and paradoxical. The preoccupation with pathological drinking—a condition that subjects the unsuspecting drinker to uncontrollable and depraved impulses—emerged in the shadows of the historical process in American culture that identified middle-class success as a lifelong moral project in which the white, male striver adhered to a strict regimen of industrious work habits, piety, and moral self-restraint. Identifying delirium tremens with failure, physicians portrayed inebriates as romantic figures struggling with their dark and diseased imaginations. Literary, poetic, theatrical, and visual representations of the disease explored and expanded on the meanings and significance of this struggle. In American popular culture, the psychic power of intoxication and the compulsive nature of heavy drinking came to dramatize fraught issues of social success and failure in a culture obsessed with both. Through delirium tremens, alcohol addiction became a psychological and physiological disease that reaffirmed middle class values and exerted a perverse fascination born of status anxiety, repression, and desire.

ONE

Ardent Spirits and Republican Medicine

Late in Charles Brockden Brown's novel *Edgar Huntly, or Memoirs of a Sleep-walker* (1799), the narrator struggles through the rugged wilderness of rural Pennsylvania trying to escape marauding bands of Indians. Terrified, starving, and shivering in soaking clothes, Edgar Huntly comes upon a stately house just as night is beginning to fall. The wooden house appears to be "the model of cleanliness and comfort . . . the abode not only of rural competence and innocence, but of some beings raised by education and fortune above the intellectual mediocrity of clowns." Hoping to dry his clothes and rest by a warm hearth, Huntly finds the kitchen door wide open. Dishes are scattered and broken and the floor is half burned by a fire that has just been extinguished.

Compelled to search the silent house, he comes upon a sleeping man. Huntly rouses him with difficulty. In a stupor, the man yells, "Is't you, Peg? Damn ye, stay away, now! I tell ye, stay away, or, by God, I will cut your throat!—I will!"

Huntly despairs: "These were the accents of drunkenness, and denoted a wild and ruffian life. They were little in unison with the external appearances of the mansion."

As he leaves the house, the nightmare continues when he comes across the drunkard's terrified wife trying desperately to quiet her crying baby: "Ah, me babe! . . . Thou art cold and I have not sufficient warmth. . . . Thy deluded father cares not if we both perish."

A few steps away, Huntly stumbles on a gruesome scene. "It was the corpse of a girl, mangled by a hatchet. Her head, gory and deprived of its locks . . . this quiet and remote habitation had been visited, in their destructive progress, by the Indians. . . . her scalp, according to their savage custom, had been torn away to be preserved as a trophy."[1]

In early national Philadelphia, religious leaders, judges, lawyers, wealthy philanthropists, hospital and almshouse administrators, poor relief advocates, and other prominent citizens dwelled increasingly on the problem of intemperance.[2] Brown's nightmare expressed the particular symbolism that drunkenness assumed in the political culture of the 1790s. American republicans subscribed to eighteenth-century theories of history holding that all societies moved through progressive stages of development. Hunting societies such as the social groups formed by American Indians, characterized by ignorance and savagery, represented the least-developed stage. Old World Europe, with its booming industry and commerce, overpopulation, and vast disparities of wealth typified the other end of this spectrum of social development. In between stood the American republic—an egalitarian society of yeoman farmers. Many believed this experimental form of government could exist only in this optimal middle ground. The primarily agricultural character of the new United States provided the social and economic conditions crucial to shaping a virtuous electorate. Access to land meant all men could live by the fruits of their own industry, elevating them above ignorance and savagery while keeping them free of corrupting luxuries.[3]

Brown's drunkard father symbolized the dangers that economic development posed to fragile virtue, embodied in the wife and children. Huntly tells us the house had been built by a farmer named Selby, "who united science and taste to the simple and laborious habits of a husbandman." The drunkard was Selby's son, who had lived for several years in Europe, where he had no doubt acquired his destructive habit. He had inherited the farm when Selby died. Brown dramatized the perversion of patriarchal authority as the incapacitated man threatens to kill his innocent wife while his daughter lies scalped by savages. We later learn that a local militia saved the father's life by driving off the Indians and dousing the house fire. In Brown's republican morality tale, habitual drunkenness linked the extreme ends of the spectrum of social development—an expression of luxury and decay that manifested itself as violent savagery.

In eighteenth-century America, as in many parts of the world and throughout history, alcohol abuse was a ubiquitous and controversial problem.[4] But how Americans perceived, defined, and responded to intemperate drinking changed markedly over the course of the century. In the years after the Revolution, leading citizens dwelled on the fragile nature of this experiment in republican government. They cited virtue, sympathy, and independence as values crucial to citizenship. Without a uniquely virtuous electorate, American freedom would be lost.[5] Heavy and compulsive drinking came to epitomize behaviors that were antithetical to these requirements. Politicians and ministers had long railed against the dangers of drink, but anti-alcohol activism took on new urgency during these years.[6]

The individuals most influential in shaping new responses to intemperate drinking were physicians.[7] The most notable temperance author and activist was Philadelphia physician Benjamin Rush. Synthesizing an existing medical literature on the health consequences of heavy drinking, Rush articulated a new argument that was appropriated by a wide range of nineteenth-century religious figures and social reformers. The growing authority of scientific medicine and new conceptions of human health transcended denominational differences and social class. Rush's pamphlet *An Inquiry into the Effects of Spirituous Liquors on the Human Body, and Their Influence on the Happiness of Society* (1784) was widely read. A scant twelve pages when it was first published, Rush revised, rewrote, and expanded his inquiry, and in 1805 he published a more substantial forty-eight-page essay titled *An Inquiry into the Effects of Ardent Spirits upon the Human Body and Mind*.[8] The broad circulation of his temperance writing made it increasingly difficult for any literate person to deny that habitual intoxication carried grave risks for body and mind. Repeatedly reprinted throughout the nineteenth century, Rush's *Inquiry* is central to his reputation among historians as the father of the American temperance movement.[9]

In shaping future American attitudes toward drinking, however, at least as important as his temperance writing was Rush's influence on the new American medical profession. He rose to be the most famous and popular professor at the University of Pennsylvania, the nation's preeminent medical school. By the end of his life, he counted over three thousand students who went on to practice medicine throughout the country. In his widely published lectures, which were standard reading for American physicians, Rush elaborated a complete system of medicine appropriate for the new republican nation. His theories describing the interdependence of physical, mental, and moral health reflected his conviction that the medical profession had a crucial role in the preservation and development of the new republic. This sense of mission would inform the emerging identity of the American medical profession. One of the most important ways physicians expressed this sense of social responsibility was through a strong commitment to temperance activism.[10]

In his medical lectures, Rush returned again and again to the consequences of heavy drinking to demonstrate his theories regarding the interdependence of physical, mental, and moral health, and he repeatedly revised his views on the nature of the problem. Late in life, he advanced the new theory that compulsive drinking constituted a physical and mental disease. Searching for the intellectual roots of modern medical theories, historians of medicine point to this move in Rush's writing as the first articulation of a "clearly developed modern conception of alcohol addiction." In his often cited essay "The Discovery of Addiction: Changing Conceptions of Habitual Drunkenness in America," sociologist Harry Levine argues that Rush was the first to describe habitual drinking as a disease characterized by loss of control.[11] Particularly, historians cite Rush's idea that habitual drinking constituted a "disease of the will" as pointing toward twentieth-century conceptions of addiction.[12]

In truth, the parallels drawn between Rush's disease of the will and what today is commonly called "alcoholism" reveal more about modern medical history than about the history of the early republic. Twentieth-century medical and social responses to problem drinking were heavily focused on addiction, often to the exclusion of broader social consequences.[13] This singular focus has also shaped recent historical inquiry.[14] Rush, like other eighteenth-century writers, did not share this preoccupation. He was broadly concerned with the political, social, economic, and health consequences of heavy drinking. Published just a year before he died, the disease of the will theory received little attention, and there is no evidence that it shaped medical practice in any significant way.[15] Nevertheless, the theory did signal a new direction in medical thinking, albeit in a negative sense. Rush articulated this theory in a defensive effort to prop up his republican system of medicine, which students and colleagues had come to see as archaic. Far from being the architect of a new theory of addiction, Rush was reluctantly trying to reconcile his political commitments and outdated medical system with the new intellectual and cultural currents sweeping the Philadelphia medical community. These currents, which Rush sought to dismiss, would subsequently shape nineteenth-century medical responses to alcohol abuse.

The course of Rush's thinking also illustrates a broader theme in the history of the early national period. In the 1780s and 1790s, reformers and politicians articulated grand hopes and aspirations for the new nation, but this republican order ran into the hard realities faced by a growing, diverse, and unruly nation.[16] In the years after the Revolution, Rush was among the most utopian of republican writers. He strove for nothing less than "to convert men into republican machines," believing "this must be done, if we expect them to perform their parts properly, in the great machine of the government of the state."[17] His teaching, his medical theories, and his social activism were imbued with this lofty mission, to perfect citizens and thereby create a harmonious and free society. Rush's failed efforts to defend the republican body from the evils of ardent spirits illustrate the stark shortcomings of his project.

A River of Death

In the eighteenth-century British Empire, the growing consumption of potent distilled liquor sparked a new medical awareness of the health consequences of heavy drinking. In England, the use of distilled liquor grew dramatically in the first half of the eighteenth century. Historians estimate that the amount the average adult drank annually increased by more than seven times between 1700 and 1743—from one-third of a gallon to 2.2 gallons.[18] This trend was due in large part to new laws that liberalized the use of distilleries and to large quantities of cheap grain, which could be converted into liquor and sold. In 1742, at the height of what became known as the gin craze, eight thousand London dram shops dispensed an estimated nineteen million gallons of gin. Rum, brandy, port, and heavy porter also grew in quantity, availability, and popularity.[19]

The key development in the history of drinking habits in British North America was the expansion of sugar plantations in the West Indies.[20] In an era when drinking water was often unreliable, colonists considered fermented beverages healthful. At daily meals they drank beer, cider, and wine at home and in colonial taverns.[21] Beginning in the last quarter of the seventeenth century, molasses brought from West Indian sugar plantations sparked a growing distilling industry centered in Philadelphia and Boston. Easily transported over long distances, rum became a staple of the booming Atlantic economy, transforming the tastes of colonial consumers.[22] For much of the century it was the drink of choice in American taverns.[23] Calling the growing availability of rum a "River of Death," the Puritan minister Cotton Mather claimed that Boston, a city of fewer than ten thousand people, imported approximately 78,750 gallons of rum each year for consumption and trade. The additional quantity distilled in the colony, he said, was unknown.[24] The Atlantic distilling industry greatly increased the quantity of liquor available to British consumers.[25]

The Revolutionary War disrupted trade with the West Indies and crippled the rum distilling industry, but this development failed to blunt Americans' love of strong drink. In the West, whiskey came to serve an important economic function. In the 1780s and 1790s, settlers poured into the trans-Appalachian frontier, but they were largely cut off from eastern markets. For many farmers the only practical means of selling their grain was to convert it into whiskey. In frontier communities, most farmers had backyard stills that could produce large quantities. By the early nineteenth century, overproduction forced the price of whiskey down to as little as thirty cents a gallon in some years. For an American public that had already developed a taste for hard liquor, ready availability and low prices drove what historians have called an American "whiskey binge" that lasted well into the century.[26]

As the growing ubiquity of liquor transformed alcohol use, perceptions of what constituted alcohol abuse also changed. Before the American Revolution, three impulses were especially important in shaping perceptions of problem drinking: protestant moralism, elite concerns about growing urban poverty, and new medical beliefs associated with the Scottish Enlightenment.

In the seventeenth century, religious leaders addressed the evils of drunkenness far more often than medical authorities did.[27] In North America, the most intense opposition to drinking was led by Puritan ministers. As New England became more socially diverse, especially the seaport of Boston, drunken citizens came to embody the growing fear that colonists had strayed from a cooperative sense of religious purpose.[28] Ministers denounced intemperance from the pulpit, and governing bodies restricted tavern licenses. Cotton Mather's father, the minister Increase Mather, wrote in *Wo to Drunkards* (1673) that drunkenness could lead to eternal damnation and dwelled on the importance of temperate consumption. "Drink is in itself a good creature of God, & to be received with thankfulness," he intoned, but "the abuse of it is from Satan: The Wine is from God, but the Drunkard is from the Devil."[29] When the sermon was republished in 1712, Mather included a preface that lamented the growing taste for rum. He warned, "Great authors have affirmed that Drunkenness has Slain more than the Sword has ever done. If only Bodies had been Destroyed by it, the Evil had not been so Woful; but it is the Ruine of Millions of immortal Souls."[30]

After the 1720s, however, efforts by religious ministers and colonial authorities to severely restrict alcohol consumption largely faded in British North America. In New England, religious warnings about drunkenness and legal efforts to regulate taverns had failed to change drinking habits. While continuing to monitor the number of taverns, colonial governments liberalized liquor laws, and the production and sale of rum became major sources of tax revenue. In Pennsylvania, for instance, authorities carefully scrutinized applicants' moral qualifications before granting tavern licenses.

But for the most part, colonial governments were complacent about restricting liquor sales.[31]

When Rush wrote his first temperance pamphlet in 1784, medical attention to alcohol abuse was a relatively recent development. Seventeenth- and eighteenth-century physicians saw wine and other fermented beverages as healthful drinks and useful medicines when taken in appropriate amounts.[32] Medical writing lumped the dangers of heavy alcohol consumption together with the consequences of overeating and rich foods, but compulsive drunkenness was not a major topic of concern. First published in 1621, Robert Burton's encyclopedic *Anatomy of Melancholy* was the most widely read work on mental illness in the seventeenth century. Burton saw gluttony and drunkenness as a cause of mental disease: "There is not so much harm proceeding from the substance itself of meat, and quality of it, in ill-dressing and preparing, as there is from the . . . intemperance, overmuch, or over little taking of it. . . . This gluttony kills more than the sword. . . . And yet for all this harm, which apparently follows surfeiting and drunkenness, see how we luxuriate and rage in this kind."[33]

But Burton did not distinguish alcohol as a distinct cause of disease, and mentions of it are few and far between. Nor did heavy drinking appear as a major concern in the practice of the eminent seventeenth-century British physician Richard Napier, whose extensive surviving records chart the multitude of physical and mental maladies plaguing his middling and elite patients.[34]

Medical concern increased along with liquor consumption. British elites became alarmed by the social consequences of the gin craze that reached its height in the 1730s and 1740s.[35] Physicians and commentators focused especially on what we would today call public health. William Hogarth's famous engraving *Gin Lane* (1751), for instance, captured elite perceptions of the consequences of gin consumption among the lower classes (fig. 1). It offered a horrific vision of the social decay caused by poverty and alcohol abuse. At the center of the painting is a drunken, syphilitic woman carelessly dropping a baby, and just below her is a horribly emaciated blind man. Explaining why he published the print, and several others with similar themes, Hogarth claimed that "the subjects of those Prints are calculated to reform some reigning Vices peculiar to the lower Class of People."[36]

Medical warnings about the dangers of drink also began to circulate in popular health books physicians wrote for an educated reading public. An important tenet of this new generation of medical writers was that Britain's growing affluence created new health risks.[37] In the influential *Essay of*

Figure 1. William Hogarth, *Gin Lane* (London, 1751).

Health and Long Life (1725), Scottish physician George Cheyne warned that "since our Wealth has increas'd and our Navigation has been extended, we have ransack'd all the Parts of the *Globe* to bring its whole Stock of Materials for *Riot, Luxury*, and to provoke *Excess*."[38] Focusing his appeal on Britain's "middling Sorts," he urged that a simple diet was the best means of preserving health.[39] Two of the most widely read health guides were the Swiss physician S. A. D. Tissot's *Advice to the People in General with regard to Their Health* and the Scottish physician William Buchan's *Domestic Medicine*.[40] Perhaps the best-selling book of its kind in the eighteenth century, Buchan's manual was reprinted in many forms in Britain and America long into the nineteenth century.[41] He devoted an entire chapter to the health consequences of "intemperance," which he defined to include the dangers associated with overeating, drinking, and sexual promiscuity: "Men . . . create artificial wants, and are perpetually in search of something that may gratify them; but imaginary wants never can be gratified. Hence, the epicure, the drunkard, and the debauchee seldom stop in their career till their money, or their constitution fails."[42] The first American temperance advocate to draw extensively on this new medical literature was the Philadelphia Quaker, teacher, and social reformer Anthony Benezet. He was a friend of Benjamin Rush, and his temperance appeals interspersed long excerpts from Cheyne, Buchan, and other physicians with religious warnings that liquor consumption led to the loss of faith.[43]

For these eighteenth-century physicians and writers, alcohol abuse was a problem as much social as medical. These authors moved quickly between the health consequences to individuals, families, communities, and nations. They portrayed alcohol consumption alternatively as a moral choice by the individual and as an expression of social circumstances. For the poor, drinking offered an escape from hardship, hunger, and despair. Intemperance among the rich derived from indolence and luxury. Merchants often gave themselves up to drink following a downturn in business. Drinking threw artisans into poverty, merchants into debt, and the rich into depravity. Authors often equated alcohol consumption with national character, citing the varied drinking habits of China, Japan, and Suriname as well as Holland, France, and Russia.[44]

Alcohol addiction was not an important subject. Eighteenth-century authors did note that the habit of drinking potent liquor often became a compulsive desire. Cheyne warned that anyone who began to crave liquor was in grave danger of falling into an inexorable downward spiral: "Drops beget Drams, and Drams beget more Drams, till they come to be without weight and without Measure; so that at last the miserable creature suffers a true Martyrdom. . . . Higher and more severe Fits of Hystericks, Tremors, and convulsions," climaxing in death.[45] Buchan warned that despair made alcohol dangerously alluring:

> The miserable fly to it for relief. It affords them indeed a temporary ease. But, alas, this solace is short-lived, and when it is over, the spirits sink. . . . Hence a repetition of the dose becomes necessary, and every dose makes way for another, till the unhappy wretch becomes a slave to the bottle, and at length falls a sacrifice to what at first perhaps was taken only as a medicine.[46]

The idea that one could become a "slave to the bottle," held particular resonance for Benezet, a tireless advocate of abolition. He echoed Buchan, writing that "the unhappy dram-drinkers are so absolutely bound in slavery to these infernal spirits, that they seem to have lost the power of delivering themselves from this worst of bondage."[47]

Benezet's equation of habitual drunkenness with slavery evokes modern understandings of alcoholism, but his meaning was quite different. Growing concerns circulating in the Anglo-British world about the medical, social, and moral consequences of alcohol abuse became politicized in the crisis that was engulfing the American colonies. In 1774, Benjamin Rush argued that excessive use of spirituous liquors would lead to indolence and moral decay. He urged, "Let the common people . . . be preserved from the effects of spirituous liquors" and described the "numerous and complicated *physical* and *moral* evils which these liquors have introduced among us" as a powerful and monstrous hydra.[48] The Revolution cemented these new meanings into American political culture. Intemperate use of ardent spirits threatened to destroy the youthful vigor and virtue of the American people, the qualities that crucially distinguished the promising new country from depraved and decrepit Europe.

Tempering Revolutionary Spirits

The derivation of the passage from Brown's *Edgar Huntly* that began this chapter illustrates how intemperate drinking took on a new symbolism. Rather than tapping an abstract argument about social development, Brown drew inspiration for the drunkard father from a personal experience, which he related to a friend in a letter. Written in 1793, six years before the publication of *Edgar Huntly*, the letter detailed a "gloomy tale" that, Brown wrote, "affords infinite subject of reflection." Five or six months earlier, a man from Ireland had moved into Brown's Philadelphia neighborhood with a wife and four daughters. Brown learned that the man had "wasted a large fortune in the most expensive and pernicious amusements" and had just "escaped from his rapacious creditors." "This wretch was the slave of drunkenness" who, at least three times each week, was "raised by intoxication into a fit of madness, and exercised the most brutal cruelties on his innocent and helpless family." The letter recounted how the drunkard habitually attacked his wife, on one occasion when she was half-naked, before finally murdering her. At the end of the letter, Brown asked his friend whether the story might form the basis of a valuable tale.[49]

In this woman's brutal slaying by a "slave of drunkenness" Brown saw a larger meaning. A novelist, historian, editor, and essayist, he was an active participant in the many debates about society and government that gripped the early republic.[50] Feminine virtue in distress and half-naked women being brutalized by villains were common in popular sentimental and gothic literature and plays. These scenes were intended to arouse powerful feelings of revulsion in cultivated readers.[51] In the superheated political rhetoric of the revolutionary crisis, feminine virtue took on compelling political meanings as well. Patriots commonly cast their struggle as a defense of the fair Lady Liberty against the brutalities of British despotism, often using the imagery of sexual violation.[52] In the years after the Revolution, the prevalent belief that a republican government could survive only with a uniquely virtuous citizenry further reinforced the powerful symbolism of vulnerable women.[53] In *Edgar Huntly*, Brown drew on these meanings to recast a tragedy he had witnessed as a horrifying vision of republican liberty at risk from the brutish drunkenness of indolent citizens. Perceptions of alcohol use and abuse came to dramatize the stark contrasts between liberty and slavery, virtue and depravity.

Rush became the most active and influential figure shaping these new perceptions. Focusing on the dangers of distilled liquor, he wrote the 1784 *Inquiry* for a popular audience, and in length, structure, and tone it closely resembled Benezet's earlier pamphlets. Little of the medical content of Rush's *Inquiry* was original. Only a page and a half of the first edition discussed the health consequences of alcohol abuse for the individual. The rest of the pamphlet laid out the pernicious influence of intemperance on the family and society. He synthesized the observations of Buchan, Cheyne, and other European writers to challenge popular beliefs, such as the common conviction that drinking spirits could be healthy in very cold or very warm weather or while performing hard labor.

What distinguished Rush's pamphlet was his argument that alcohol abuse posed a dire threat to liberty. After listing the many adverse health consequences of drinking, he wrote, "A people corrupted with strong drink cannot long be a free people." If the nation's elected leaders reflected the will of the people, "all our laws and governments will sooner or later bear the same marks of the effects of spirituous liquors, which were described formerly upon individuals."[54] In a later edition of the *Inquiry* he revised this passage to read, "The customs of civilized life . . . cannot prevent our country from being governed by men, chosen by intemperate and corrupted voters. From such legislators, the republic would soon be in danger.[55] The fragility of the republic demanded that citizens maintain healthy habits.

Why did Rush focus so much energy on temperance, as opposed to many other challenges facing the young nation? Throughout history and in various countries, concerns about alcohol or drug abuse have often coincided with new consumption patterns, as with the growing availability of distilled liquor in colonial Massachusetts.[56] But in 1784, when Rush's *Inquiry* was published, consumption of distilled liquor was probably not appreciably greater than in earlier decades and may have been lower, since the Revolution had disrupted the rum-distilling industry. Rush's campaign also began almost a decade before violent opposition to Alexander

Hamilton's excise tax on whiskey engulfed western Pennsylvania. Further, taverns had played an important role in the Revolution, a cause to which Rush had devoted his life. Popular places for patriots to gather, taverns had been viewed by the British as hotbeds of sedition.[57] Why then did Rush initiate a campaign against American drinking culture when taverns had provided nourishment for republican sentiment to blossom?

Part of the answer is that Rush began his crusade at a moment when many prominent citizens believed the Revolution was under threat from public disorder.[58] Social upheaval, economic hardship, and political uncertainty marked the 1780s. Bitter political debates raged over the limits of popular representation in new state constitutions. The Continental Congress's war spending and fiscal policies created an economic depression in 1784. Especially in rural areas, hardship caused by the dislocations of the war and wild inflation created discontent with new state governments. In many rural areas, debt-ridden farmers violently resisted tax collectors and creditors.[59] In Philadelphia, leading newspapers and magazines regularly printed essays decrying the rise of poverty, criminality, and immorality. Anxiety particularly focused on the drinking habits of the poor and on public drunkenness around taverns and fairs.[60] By 1787, popular unrest and economic crisis motivated some governing elites to call for a constitutional convention and a more muscular federal government.

Rush shared these elite anxieties. During the ratification debate, he supported the newly drafted Constitution and the strong federal government it proposed. Evidence suggests that these concerns also inspired Rush's temperance campaign. He wrote the *Inquiry* after a ten-day pleasure trip through the Pennsylvania backcountry, the same area where the fictional Edgar Huntly stumbled on the drunkard father. Accompanied by a servant, Rush traveled as a gentleman, treating "himself to all the comforts that circumstances would allow."[61] During the trip, he was appalled by the presence of stills, especially on Scotch-Irish farms. He wrote in his diary, "The quantity of rye destroyed and of whisky drunk in these places is immense and its effects upon their industry, health, and morals are terrible."[62] Coming from an elite gentleman traveling for pleasure, Rush's reaction to the drinking habits of western farmers illustrates the class tensions that shaped politics in the 1780s and 1790s.

Rush's temperance campaign also had roots in his professional ambitions. In the 1780s he was seeking to cement his position within the young American medical community. He had entered the profession in the 1760s as an apprentice to a successful Philadelphia physician. These were years when the orthodox medical profession was just taking shape around new medical schools in Philadelphia and New York. The most prominent physicians and teachers in the North American colonies were men who had trained at Edinburgh, the preeminent medical school in the British Empire and the epicenter of the Scottish Enlightenment. Some three hundred men from British North America attended Edinburgh in the second half of the eighteenth century.[63] Writing on the dangers of ardent spirits was part of Rush's larger effort to secure an important place in medicine, demonstrate the utility of his medical theories, and boost the social prominence of the nascent American medical profession.[64]

When Rush journeyed to Edinburgh in 1766 to further his education, the principles of Scottish medicine were already well established in America. The several years he spent in Britain brought him into contact with the leading thinkers of his day, the most important being the physician William Cullen.[65] When he returned, Rush hitched his medical career and social standing to the Revolution. He signed the Declaration of Independence, attended the Continental Congress, and served as a surgeon general during the war. After the Revolution, Rush continued to devote his life to the development of the new republic. Speaking to a Philadelphia audience in 1787, Rush declared,

> The American war is over: but this is far from being the case with the American revolution. . . . It remains yet to establish and perfect our new forms of government; and to prepare the principles, morals and manners of our citizens for these forms of government, after they are established and brought to perfection.[66]

Writing tirelessly on temperance, abolition, education reform, penal reform, and health reform, Rush hoped to establish the foundations for a new social order that he believed would foster an enlightened citizenry.

The Republican Body

The same political commitments that inspired Rush's temperance activism also shaped his ideas about medicine and medical practice.[67] In 1790 Rush became the most distinguished teacher at the University of Pennsylvania's Medical School when he assumed the professorship of the Institutes and Practice of Medicine, a position of central importance at the university. His new role was to provide students with a complete system of medicine, and Rush used the opportunity to develop and teach a medicine that he saw as compatible with republican government. He later wrote that the impetus for this project was "the activity induced in my faculties by the evolution of my republican principles by . . . the American Revolution."[68] He held the professorship for twenty-two years, with several hundred students enrolling annually.[69] The consequences of intemperance constituted a regular subject of his lectures, illustrating the particular intellectual influences and political motives that inspired the American medical profession's lasting interest in the health consequences of intemperance.

Crucial to Rush's system of medicine, and his warnings about the dangers of drink, was the understanding that the human nervous system was susceptible to external stimulation, a property called sensibility. A pervasive subject of inquiry in the Scottish Enlightenment, the foundational text for theories of sensibility was John Locke's *Essay concerning Human Understanding* (1690).[70] Stated simply, Locke reasoned that at birth the human mind was a tabula rasa, or blank slate. All that the individual was, all knowledge, derived from interaction with the outside world. The human senses, both internal and external, gathered distinct, individual impressions and bound them together in the mind to form complex ideas. For Locke this process produced all forms of human knowledge, including morality. Writing in the mid-eighteenth century, the philosopher David Hume expanded on this concept by describing the self as a bundle of impressions provided by the internal senses and bound together by the mental faculty of imagination.[71]

Rush's formulation of the principle of sensibility was deeply indebted to the innovations of William Cullen, whose teaching at Edinburgh influenced a generation of medical practitioners. In his lectures at the University of Pennsylvania, Rush described sensation as the basis of all aspects of physical and intellectual life. He reduced the fundamental functioning of the body to the principles of "*sensation, motion, and thought*."[72] All parts of the human body had sensibility—the ability to receive impressions from the outside world—or irritability, the capacity of communicating those impressions to other parts of the body or mind. Human life depended on the constant "action of *stimuli* upon organs of sense and motion." Individuals were in constant communication with the surrounding environment, and that environment shaped and molded them both physically and mentally.

> Yes, Gentlemen, the action of the brain, the contraction of every muscular fibre, the diastole and systole of the heart, the pulsation of the arteries, the peristaltic motion of the bowels, the absorbing power of the lymphatics, secretion and excretion, hearing, seeing, smelling, taste and the sense of touch, nay more, thought itself; all depend upon the action of *stimuli* upon organs of sense and motion.[73]

These stimuli could be either external or internal. As external Rush listed "light, sound, odors, heat, pure air, and the reflected stimulus of exercise." The internal stimuli were "food, drinks, chyle, the blood, a certain tension of the glands which contain secreted liquors." He also described stimuli emanating from the mind, including "the reflected exercises of the *understanding* and of certain *passions of the mind,*" including hope, love, joy, and ambition, but also avarice, anger, fear, hatred, malice, and envy, among others.[74]

The principle of sensibility had important social implications. Entirely activated by stimuli, a person's social and natural environment, habits, diet, and education were the crucial determining factors in individual development. Rush's articulation of the principle of "sympathy" further illustrated these concerns. The whole human body was connected, he wrote, so that "impressions made . . . upon one part, excite sensation, or motion, or both, in every other part of the body."[75] Most often the principle of sympathy worked by transmitting impressions through the nervous system, but it also worked through the muscles and blood. By this principle, disease manifested itself in the body, creating negative sympathies that did not exist in health. "Vomiting," for instance, "gives us a notice of a stone in the Kidneys, and a pain in the shoulder indicates a stone in the liver." Sympathy enabled him to explain mental functions as well. Knowledge, he argued, was the product of individual sensations associated in the mind into complex forms through the principle of sympathy.[76]

Rush's understanding of sympathy and sensibility lay at the heart of his conviction that medicine had a crucial role in the social development of the new nation. This impulse to generalize medical science to the project of nation building was entwined with Enlightenment inquiry in general. Writing about the time Rush journeyed to Edinburgh, giants of the Scottish Enlightenment like David Hume and Adam Smith placed sympathy and sensibility at the center of theories regarding social solidarity and social development. These and other theorists believed human sensibility brought men together into societies. Through proper cultivation, a people's sensibility grew more responsive as nations became more civilized.[77]

To fully realize the political potential of medical science, Rush departed from his teacher Cullen in applying the principles of sensibility to the science of the mind.[78] In this he drew heavily on the Scottish physician and philosopher David Hartley, who had developed a complete physiology of the mind and soul based on the principle of sensibility.[79] Rush described the mind as made up of distinct "faculties or capacities," which he listed as instinct, memory, imagination, understanding, will, passions, the principle of faith, and the moral faculties, divided into conscience, the sense of deity, and the moral faculty. Rush also called these faculties "internal senses," because they "are in every respect like the external senses in their origin and offices." The internal senses, "are all awakened by impression, and their operations are as much the effects of specific motions, as the operations of the senses of touch, taste, smelling, seeing, and hearing." Like organs or muscles, the mind's faculties were strengthened through exercise and diminished by neglect.[80] Rush believed that liberty, for example, would reshape the minds of citizens.

> From a strict attention to the state of mind in this country, before the year 1774, and at the present time, I am satisfied, the ratio of intellect is as twenty are to one, and of knowledge, as an hundred are to one, in these states, compared with what they were before the American revolution.[81]

For Rush, popular elections themselves benefited the mental faculties of all citizens.

Nothing was more important for the survival of the nation than the moral development of its citizens. "Virtue," Rush wrote, "is the living principle of a republic."[82] In 1786 he argued to the American Philosophical Society, "As Sensibility is the avenue to the moral faculty, every thing which tends to diminish it, tends also, to injure morals." Diet, habits, and repetition, as well as music, climate, hunger, forms of government, and even odors affected the individual's capacity for sensibility. Hence Rush believed criminal punishments should be hidden from public view, lest the constant sight of violence lessen "the natural horror which all crimes at first excite in the human mind." Being habitually cruel to animals while young could become an avenue to committing murder in adulthood, because the habit of cruelty destroys moral sensibility.[83]

Because moral sensibility conformed to physical rules, man's perfection could be achieved through a proper understanding and application of their principles. In his introduction to his extensive lectures on the mind, Rush wrote that knowledge of the human mind was the most important, certain, intelligible, and above all "*most useful* of all the sciences. It is interesting to the divine, the statesman, the philosopher, the scholar, and to all persons who have anything to do with the duties, the government, the interests, the health and the happiness of man."[84] Through improved diet, habits, and education, the new nation could grow a virtuous citizenry. By applying the correct stimuli to the electorate, the principles of sensibility would inevitably shape a virtuous nation.[85] These laws also implied that citizens had to be constantly protected from adverse sensations. Man was both perfectible and profoundly vulnerable.

Rush's understanding of disease was also inseparable from his belief in republican theories of social development. He argued that all diseases, physical and mental, derived from a single cause: morbid impressions, or movement, in the vascular system. At their root, diseases involved excess motion in the blood, and as such all diseases were variations of fever: "the blood vessels are the outposts of the system. . . . they receive the first attacks of morbid impressions which they discover in the different forms of fevers."[86] Fevers were also "the diseases of the first and most simple states of society." In overly civilized Europe, the essential nature of disease had become obscured. With luxury and civilization, "diseases spread to other bodily systems, particularly to the nerves, muscles, brain and mind."[87] These were chronic, or "artificial" diseases that manifested themselves in paroxysms that recurred over a long period.[88] These theories shaped Rush's controversial emphasis on bloodletting as a form of therapy. By taking blood, he believed, the doctor depleted the activity in the vascular system, draining it of the morbid impressions that were causing illness.[89] Even in treating "artificial" diseases such as gout, apoplexy, and insanity, he believed that bleeding, albeit often in combination with other therapies, remained the best way to treat the underlying cause. "Reject the unity of diseases," he told his students, "and our science becomes a mere chaos, a farrago of unmeaning words, and a compound of folly & ignorance."[90]

Liquor offered Rush compelling demonstrations of his medical theories and their social implications. Drunkenness dramatized the susceptibility of the moral faculty to diet, habit, and other physical influences. In the early *Inquiry*, Rush wrote that the effects of liquor on the moral faculty were "distressing and terrible." Spirituous liquors altered moral behavior and judgment, making men "peevish and quarrelsome. . . . They violate promises and engagements without shame or remorse."[91] Ardent spirits thus demonstrated that the moral faculty could be depraved through habit and diet. Rush also constantly referred to ardent spirits in demonstrating his theory of disease. Because spirits were a powerful stimulant, they quickened the circulation of blood, producing heat in the body, but this heat quickly dissipated, draining the vital powers and producing weakness. In particular, Rush dwelled on the role of ardent spirits in causing "artificial" diseases like apoplexy, palsy, coma, convulsions, and epilepsy (all of which, he argued, were different names for one disease) as well as gout, dyspepsia, and madness. "The first use of these baneful liquors is happily characterized by the fable of Prometheus, who is said to have stolen fire from heaven,"

Rush said in his lecture on diseases of the liver: "Their effects are as happily characterized by the punishment of this theft. It was a vulture preying upon his liver."[92]

The consequences of intemperance thus offered powerful confirmation of Rush's conception of the republican body. In the 1790 edition of the *Inquiry* he published a moral thermometer, which captured his conception of the nature of intemperance as a medical, social, and political problem (fig. 2). At the top of the thermometer, epitomizing temperance, was water, followed closely by milk and water, vinegar and water, molasses and water, and small beer. Next to these followed the positive effects of these drinks, including health, wealth, serenity of mind, reputation, long life, and happiness. Under the first grouping was a second grouping of fermented drinks, including cider, wine, and porter. This second group promised cheerfulness, strength and nourishment, but "when only taken at meals, and in moderate quantities."

Although Rush was not the first to publish a moral thermometer, he was the first to use it as a metaphor for intemperance.[93] The lower part of the thermometer marked the descent into intemperance. Weak punch was just above the midline, but strong punch fell into a group with toddy, grog, flip, and sling. At the bottom, the worst drinks included, in order, bitters infused in spirits, morning drams, and pepper in rum. Next to these drinks were separate categories of vices, diseases, and punishments. Vices ran from idleness and peevishness up by "toddy," to the vices associated with morning drams, which included "hatred of just government." The lowest vices were murder and suicide. Similarly with diseases, drinkers of strong punch and toddy risked gout, sickness, and puking, but drinking rum promised madness, palsy, apoplexy, and death. The scale of punishments ran from debt to the gallows. Like the mercury in the weatherglass, so too with the habit of intemperance—the impulse moves up and down seemingly of its own volition but according to fixed physical principles. Illustrating his more general view that man's moral virtue was shaped by sensation and obedience to natural laws, the thermometer captured the danger liquor posed to the ideal republican citizen.

Always intended for a popular audience, the thermometer was printed and reprinted by temperance activists throughout the nineteenth century. Rush's medical writing on intemperance, however, increasingly departed from this analogy. After 1800, he began to grapple with a difficult question that the thermometer left mysterious: Why do people continue to drink despite the horrible consequences? This question had troubling implications, not just for his temperance campaign but also for his entire system

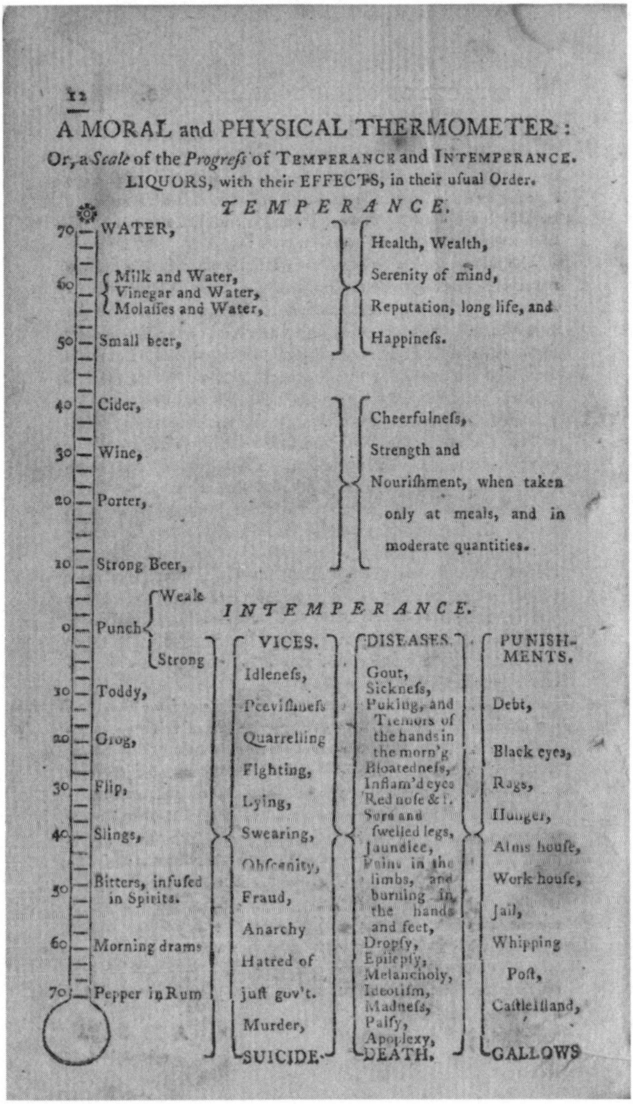

A MORAL and PHYSICAL THERMOMETER:

Or, a *Scale* of the *Progreſs* of TEMPERANCE and INTEMPERANCE.

LIQUORS, with their EFFECTS, in their uſual Order.

TEMPERANCE.

Scale	Liquors	Effects
70	WATER,	Health, Wealth, Serenity of mind, Reputation, long life, and Happineſs.
60	Milk and Water, Vinegar and Water, Molaſſes and Water,	
50	Small beer,	
40	Cider,	Cheerfulneſs, Strength and Nouriſhment, when taken only at meals, and in moderate quantities.
30	Wine,	
20	Porter,	
10	Strong Beer,	

INTEMPERANCE.

Scale	Liquors
0	Punch (Weak / Strong)
10	Toddy,
20	Grog,
30	Flip,
40	Slings,
50	Bitters, infuſed in Spirits.
60	Morning drams
70	Pepper in Rum

VICES.
Idleneſs,
Peeviſhneſs
Quarrelling
Fighting,
Lying,
Swearing,
Obſcenity,
Fraud,
Anarchy
Hatred of juſt gov't.
Murder,
SUICIDE.

DISEASES.
Gout,
Sickneſs,
Puking, and Tremors of the hands in the morn'g
Bloatedneſs,
Inflam'd eyes
Red noſe & f.
Sore and ſwelled legs,
Jaundice,
Pains in the limbs, and burning in the hands and feet,
Dropſy,
Epilepſy,
Melancholy,
Ideotiſm,
Madneſs,
Palſy,
Apoplexy,
DEATH.

PUNISHMENTS.
Debt,
Black eyes,
Rags,
Hunger,
Alms houſe,
Work houſe,
Jail,
Whipping Poſt,
Caſtle Iſland,
GALLOWS

Figure 2. Benjamin Rush, "A Moral and Physical Thermometer," in *Inquiry into the Effects of Spirituous Liquors on the Human Body* (Philadelphia, 1790).

of republican medicine. What if moral development did not conform to natural laws? Why do republican citizens persist in irrational behaviors that threaten to destroy the nation?

His struggle to answer these questions occurred as part of his more general ambivalence about the political and social development of the new nation. He supported the new Constitution ratified in 1789 but was soon appalled by the elitist behavior and conservative policies of George Washington's Federalist administration. Joining the Democratic Republicans, he believed the Federalists were attempting to establish a European-style aristocracy antithetical to the egalitarian spirit of 1776. While he greeted Thomas Jefferson's election with hope, he remained increasingly pessimistic about the direction of the republic. At the same time, intellectual developments rendered his medical system increasingly antiquated. After 1800, the direction of medical inquiry challenged Rush's views on disease, physiology, and psychiatry. In the last ten years of his life he published numerous works defending his vision of republican society and his system of medicine. Even so, these later writings grew increasingly "gloomy" and self-indulgent as he slowly retreated from his views on the radical perfectibility of mankind.[94]

Nowhere was this pessimism more apparent than in Rush's writing on intemperance. In a letter to his longtime correspondent John Adams in 1808, Rush related an elaborate dream he had the night after losing a patient to "the fatal effects of ardent spirits." Rush went to sleep contemplating his patient's death and dreamed he had been elected president of the United States. He quickly banned the import, manufacture, and sale of all ardent spirits. "Wise, humane, and patriotic as this law was," however, "it instantly met with great opposition." Critics argued that the ban threatened to wreck the economy. Farmers, lawyers, clergyman, cab drivers, and tavern keepers needed liquor to work. President Rush objected, "You don't know the people of the United States as well as I do; they will submit to the empire of Reason, and Reason will soon reconcile them to the restrictions and privations of the law for sobering and moralizing our citizens." But President Rush ultimately conceded that citizens would never abide by the liquor ban. As a wise councillor explained, "Mr. President, in thus rejecting the empire of Reason in government, permit me to mention an empire of another kind, to which men everywhere yield a willing, and in some instances involuntary, submission, and that is the empire of Habit." President Rush was unceremoniously escorted from his office and advised "to go back to your professor's chair and amuse your boys with your idle and impracticable speculations."[95] Written when he was sixty-two, and perhaps feeling his age, Rush's account of the dream reflects despair over the failure of his temperance campaign.[96] Although Rush continued to reprint his temperance pamphlet until the end of his life, the wise councillor's admonishment that "you might as well arrest the orbs of heaven in their course as *suddenly* change the habits of a whole people" reads like an epitaph to Rush's naive belief that exposing the evils of liquor would naturally lead republican citizens to drink less.

A Disease of the Will

Rush's medical writing reflected the same frustration apparent in his dream. Whereas previously he had focused on alcohol as a cause of disease, after 1800 he increasingly investigated drinking as a disease in itself. In one lecture, for instance, Rush presented the disease of drunkenness as chronic apoplectic paroxysms:

> These, when they occur in company and in a tavern, singing, hallowing soaring, imitating the noises of brute animals, jumping, tearing off clothes, and dancing naked. Breaking glasses, throwing bottles at the heads of waiters and cooking gold watches in with hogs lard in a frying pan. This [gentlemen] is a picture drawn from the life of scenes which have occurred within doors in this city. Sometimes the Insanity produced by strong drink discovers itself by a company this deranged, rushing out after midnight into the streets, filling the air with their yells and bacchanalian songs, tearing down and misplacing signs, breaking of knockers and insulting and knocking down watchmen. Well would be for these Belials if their follies & vices ended here. From the acts of violence which have been mentioned they proceed to houses of ill fame where they pass the remaining part of the night in the grossest acts of debauchery and rioting, alternately flattering and cursing, caressing & kicking, the gals and Pegs, and Bettys and Kittys of the town, till overcome by strong drink they pass on to the last effects produced by it, that is apoplectic drunkenness.[97]

Here he did not separate drunkenness from the habit of drinking liquor. Drunkenness was a chronic, sometimes hereditary, and perhaps contagious disease that was manifested in paroxysms that grew ever more violent, finally recurring "every hour of every day." This description also remained rooted in eighteenth-century theories of the mind and body. Rush portrayed the pathology of drunkenness as a mechanistic function of the mind, saying that it lay in "the *association of ideas*."[98] He recommended subjecting the drunkard to extreme humiliation or fright, reasoning that the pleasurable feelings that accompanied drinking could be replaced with a very negative association. Religion proved the best cure, however: "If Christianity had nothing else to recommend it but superior and almost exclusive curing of this destructive disease of the body and mind, it would be sufficient to entitle its doctrines to our belief, and its duties to our practice."[99]

Rush saw no contradiction in calling drunkenness a "physical vice," a paroxysm caused by apoplexy, and a moral failing—the word "disease" applied to all these concepts.

In the last few years of his life, however, Rush came to the conclusion that the roots of the disease of drunkenness ran far deeper—to the core of the republican being. In *Observations on Diseases of the Mind* (1812), he reasoned that habitual drinking could be a symptom of "moral derangement" that manifested itself as a "disease of the will." The "symptoms of this disease" were acts of radical evil, depravity, and criminality, including especially murder, theft, lying, and habitual drinking. Rush wrote:

> An attachment to *strong drink* is at first the effect of free agency. But from habit it takes place from necessity. . . . That persons who are devoted to strong drink, act from necessity I infer from their being [irredeemable] by all the considerations which domestic assertion, friendship, reason—interest, reputation, property, and even Religion can suggest to them.[100]

In the 1784 edition of the *Inquiry*, Rush had made no mention of compulsive drinking, confident that republican citizens could be formed and reformed. Here he portrayed the compulsion to drink as overwhelming, depraved, and evil. He recommended that, after evaluation by a court composed of two judges and a physician, hard drinkers be forcibly confined in a "SOBER HOUSE" built exclusively for their incarceration. While inebriates "are as much objects of public humanity and charity, as mad people," he wrote,

> they are indeed more hurtful to society, than most of the deranged patients of a common hospital would be, if they were set at liberty. Who can calculate the extensive influence of a drunken husband or wife upon the property and morals of their families, and of the waste of the former, and corruption of the latter, upon the order and happiness of society?[101]

In his mind, heavy drinking had become just one symptom of a larger disease that also manifested itself as pathological murder.

Why did Rush's thinking take such a hard turn? Frustration with the persistence of drinking in the new republic is only part of the answer. His preoccupation with moral derangement derived from a much larger intellectual endeavor. After 1808, Rush began to deliver a new series of lectures to his students directly addressing a wave of innovative European writing on psychology that had begun with Erasmus Darwin's *Zoonomia, or The Laws of Organic Life in Three Parts* (1793).[102] He later published these lectures as *Observations on Diseases of the Mind*. The first American book-length exploration of mental illness, the book earned Rush the reputation among modern historians of medicine of the father of American psychiatry.[103] That reputation has tended to obscure the fact that *Observations* constituted a significant intellectual departure from his previous subjects. Previously, mental disease had represented the type of "artificial" nervous disorders that Rush associated with older European societies, not youthful America. In this project he sought to reconcile his system of medicine and his political commitments with the new direction of medical science.

This outpouring of European writing reflected two interrelated developments. First, a generation of physicians developed and expanded on new models of human physiology. Many of the most important writers followed Cullen in exploring the role of the nervous system in both physical and mental disease. Like Rush, many of these British writers, including Thomas Arnold, Alexander Crichton, John Ferriar, William S. Hallaran, and Thomas Trotter, were former students of William Cullen. The works of other British writers, including Joseph Mason Cox and John Haslam, as well as German and French authors, most significantly the Parisian Phillppe Pinel, also circulated in the Philadelphia medical community through institutions such as the Pennsylvania Hospital library and the Library Company of Philadelphia. Booksellers regularly advertised medical treatises in newspapers.[104] The second development was the emergence in England of a growing number of mental asylums overseen by physicians. Asylums provided doctors with large concentrations of mentally ill patients that they could observe and study over a long period.[105] These physicians attempted to reconcile Cullen's theories with mental illness.[106] Often differing in their terminology, models of the mind, and theories of disease, these authors all shared assumptions about the mind and body that were deeply indebted to Lockean sensationalism.

In focusing on insanity, altered mental states, and extreme emotions, these authors described a mental life that challenged the eighteenth-century views that had influenced Rush in his early lectures. Taken together, these authors began to articulate a new brain science that departed from the mechanistic models of Locke and Hartley by describing the mind in increasingly biological terms.[107] One of the most important lines of inquiry in this new science was the relation between mental states and physical structures. In the seventeenth century, Locke had argued that madness was the result of an imperfect association of ideas. Locke wrote, "Madnesse seems to be noething but a disorder in the imagination."[108] The implication was that insanity was a primarily mental condition and a departure from the normal workings of the faculties of the mind. But writing in 1801, for instance, Pinel included in his influential treatise on mental illness an extensive section on craniology, comparing the dimensions of his patients' skulls with ideal types:

> I have examined the relation of the height of different skulls, with their depth in the direction of the great axis of the cranium, and with their breadth at the anterior and posterior part of the same horizon. I have marked the want of symmetry in the corresponding parts, and compared, in the living subject, the bulk of the head, or rather its perpendicular height with that of the whole stature. In order to attain to some degree of accuracy in my investigation, I have taken for my standard, the admirable proportions of the head of the Apollo, as they are given by Gerard Audran.[109]

Pathological mental states thus came to suggest physical deviance from ideal norms.

These assertions suggested a theory of body and mind markedly different from Rush's view of the republican citizen as formed and activated entirely by external stimulation. A new model of psychological interiority emerged, as these authors posited that some aspects of mental life were innate—not formed by sensation. In *An Inquiry into the Nature and Origin of Mental Derangement* (1798), the Scottish physician Alexander Crichton wrote, "Man is not a self-active being whose conduct depends intirely [*sic*] on impulses which originate within himself." Nevertheless, within the psyche lay potentials or inclinations, affirming for Crichton that man "may be said to contain within himself the secret springs of his own conduct."[110] The new brain science, then, gave rise to increasing speculation on the nature of these "secret springs" and shaped a sense in Western culture that each individual has a unique inner mental life.[111]

Rush's theory that habitual drinking constituted a disease of the will has to be understood as an expression of larger concerns. Tracing its development demonstrates that Rush was both pushing against these new conceptions of the mind, which were beginning to have a profound influence on

American medicine and culture, while also contemplating the social development of the early republic. Heavy drinking was not the initial inspiration for Rush's theory of "moral derangement," or even his main concern. He first described moral derangement as he contemplated the more general problem of radical human evil. It was only after he developed the disease model that he included drinking as a symptom.

Rush first came to the idea in revisions to a lecture titled "Facts and Documents on Moral Derangement as Exemplified Chiefly in Murder," written sometime in 1809 or 1810. He had long believed that the moral faculty could become depraved, but this lecture significantly expanded his views on the subject as he developed his argument that the act of murder could sometimes be a symptom of a mental and physical disease.[112] Preserved by the Library Company of Philadelphia, these lecture notes show considerable revision. Preserved manuscripts demonstrate that Rush commonly reworked lectures many times, but this set of notes is particularly messy, with sections crossed out, hurried thoughts written in the margins, and long sections added on the backs of the earlier drafts. It was in these marginal notes that Rush first wrote down the idea that habitual drinking could be a symptom of moral derangement.

The notes record the source of Rush's preoccupation that led him to develop the disease of the will theory in the first place. Pasted into the early pages are six newspaper clippings describing sensational murder cases in great detail. The notes indicate that Rush read these news accounts to his students at the beginning of his lecture. "*Most horrid Murder!*" the first article begins: "It falls very unfortunately to our lot to communicate one of the most barbarous and murderous acts ever committed by a monster in human shape." The account narrates in bloody detail how, without warning, a man savagely attacked his parents with an ax. After dismembering his father, he turned on his mother. "Taking her bowels, heart, and liver out," he threw them in the oven, "which had just before been heated by the family to bake bread." The next day, when confronted by his brother and a group of neighbors, "the monster, after having thrown away his deadly weapon surrendered himself saying, 'I am the person who has done all this.'"[113] Apparently fascinated by murder, Rush told his students these few selections were from the "many I have met with in the course of my reading."[114]

Why would Rush include sensational newspaper accounts of highly unusual murders in a formal lecture? The notes presented no evidence that incidents of murder were rising or that they were a new and pressing threat to social order. Rather, he was responding to a question that haunted the early American republic—how to reconcile his optimistic republican view of human nature with the existence of evil.[115] Graphic accounts of murder were a new topic in early national newspapers and magazines. As described by historian Karen Halttunen, these accounts represented a new, nonfictional literature on murder that employed conventions with close parallels to the genre of gothic fiction, which swept the American book market during this period. Halttunen argues that murder became a compelling problem in the early republic because fundamentally optimistic Enlightenment theories of psychology could not explain radical human evil. As a result, these nonfiction gothic narratives of murder emphasized "its intrinsic unknowability—and its fundamental horror—the inhuman nature of the act."[116]

This preoccupation transcended Philadelphia's sharp political divisions. The writer who most thoroughly explored the troubling questions raised by murder, evil, and the irrational was Charles Brockden Brown. Although Brown moved in Federalist circles, Rush nevertheless greatly admired his work, describing him as an "eloquent" writer with a "masterly pen."[117] Like Rush's lecture on murder, Brown's novel *Wieland* (1798) also drew inspiration from graphic accounts of a sensational murder. Published in 1796 in the *Philadelphia Minerva*, the report described the murderer as a loving father who came under the influence of disembodied voices that led him to slaughter his innocent family. The editor concluded the story by stating, "The cause for this wonderfully cruel proceeding is beyond the conception of human beings" and wondered if the man acted from "the effect of insanity" or "under the strong delusion of Satan." Drawing from the new European medical literature on extreme mental states, Brown used this murder to explore the nature of mental life. In *Wieland*'s advertisement, Brown echoed the British physician Alexander Crichton, whose treatise had just been published in Philadelphia. Brown wrote, "Some readers may think the conduct of the younger Wieland impossible. In support of its possibility the Writer must appeal to Physicians and to men conversant with the latent springs and occasional perversions of the human mind."[118] Specifically, Brown used the gruesome murder to explore the frailties implied by medical theories of sensibility.[119]

The central question that *Wieland* pressed was, If sensory impressions produce all knowledge, then what happens when the senses themselves are tricked or depraved? When the father, Wieland, first hears voices, for instance, his sister Clara worries that the "effect upon my brother's imagination was of chief moment," and that it reflected "a diseased condition of his frame." "The will is the tool of the understanding," she continues, "which must fashion its conclusions on the notices of the sense. If the senses be depraved, it is impossible to calculate the evils that may flow from the consequent deductions of the understanding."[120] Throughout, Brown holds Enlightenment faculty psychology up against supernatural explanations for gothic horror, conjuring nightmarish "phantoms" and "spirits" of the mind. When Clara spends a terrifying night anxious about the well-being of a friend, she says, "Thus was I tormented by phantoms of my own creation." Hearing voices in her closet, she begins to imagine that her brother may be plotting to murder her. She is horrified not only by that possibility, but also that she could even have such a thought. Brown seems to be speaking directly to Rush and other theorists of sensibility when Clara says, "Ideas exist in our minds that can be accounted for by no established laws."[121]

Rush's lecture on murder and moral derangement responded to this challenge in two main ways. First, he sought to reconcile his conception of the moral faculty with gothic murder narratives. And second, he acknowledged new developments in European medicine while defending the principles of his own system. Addressing his students, Rush read his newspaper clippings as a way into a discussion of the relation between the moral faculty and the will, and specifically the question whether the will operates "freely" or "by necessity." His answer was both: that the will did operate freely except when affected by disease. This disease had both physical and moral dimensions. "When the will becomes the involuntary vehicle of vicious actions . . . I have called it *moral derangement*."[122] Because Rush believed that faculties resembled muscles, it followed that they could become diseased: "Exactly the same thing takes place in this disease of the will, that occurs when the arm or foot is moved without an art of the will, and even in spite of it."[123] A diseased moral faculty willed the individual to acts of radical evil.

Rush properly cited Pinel as the first writer to publish a description of the disease of the will. Pinel termed the disease a "lesion of the will" and offered a case history in his *Treatise on Insanity*, which Rush read in his lecture on murder:

> The memory, the imagination, and the judgment of this unfortunate man were perfectly sound. He declared to me, very solemnly, during his confinement, that the murderous impulse, however unaccountable it might appear, was in no degree obedient to his will; and that it once had sought to violate the nearest relationship he had in the world, and to bury in blood the tenderest sympathies of his soul.[124]

Rush departed from Pinel in asserting that the will and other faculties of the mind were akin to muscles, and he explained that the disease of the will was patterned after "every respect of a muscular disease." His proposed therapies followed his broader theories regarding the nature of disease: "For derangement of the will, no mental remedies are of sufficient force without the aid of bleeding, purging and low diet."[125] He thus appropriated Pinel's disease and fitted it awkwardly within his own theory, which described all diseases as rooted in the vascular system. For Rush, murder might be just another symptom of fever, which should be treated with the lancet.

This formulation of moral derangement became Rush's paradigm for understanding heavy drinking as one symptom of a pathological condition rooted in the brain. Scholars have generally cited Rush's proposal to treat this condition by forcibly committing drunkards to "sober homes" as a benevolent impulse, ignoring the punitive nature of his proposal.[126] Akin to murderers and worse than thieves, drunkards had to be incarcerated to protect his impressionable republican society. Straining to justify his proposal to imprison persons against their will and without a trial by jury, Rush wrote:

> Let it not be said, that confining such persons in a hospital would be an infringement upon personal liberty, incompatible with the freedom of our governments. We do not use this argument when we confine a thief in jail, and yet, taking the aggregate evil of the greater number of drunkards than thieves into consideration, and the greater evils which the influence of their immoral example and conduct introduce into society than stealing, it must be obvious that the safety and prosperity of a community will be more promoted by confining them than a common thief.[127]

Yet even thieves enjoyed constitutional guarantees for due process and the privilege of habeas corpus. Rush had come to the conclusion that drunkards were incapable of citizenship and thus undeserving even of basic civil rights. His proposed sober home was more prison than asylum.

After the work's publication, the medical community showed indifference to the disease of the will diagnosis. When Rush compiled and published his lectures as *Observations on Diseases of the Mind* in 1812, bookseller advertisements promoted the theories on diseases of the will and other mental faculties as key to the book's significance. In a published review, however, physician George Hayward dismissed them as "rarely, if ever, subjects of medical treatment" and thus unworthy even of description.[128] The proposal to build a sober home also fell on deaf ears. By the time Rush died in 1813, prominent physicians viewed his theories as archaic.[129] His writing on moral derangement and, more broadly, *Observations on Diseases of the Mind* were rearguard actions, an effort to rescue his system of medicine as his prestige in the profession declined.[130]

Perversions of the Human Mind

And yet in the years immediately following Rush's death, American physicians became far more involved in the social response to alcohol abuse than they had been during his lifetime. This involvement was not sparked by a feverish republican imperative to mold a virtuous electorate. Beyond rejecting Rush's disease of the will, Philadelphia physicians departed from his model of the republican self of sensibility and virtue—the view that man's physical, intellectual, and moral development is entirely activated and shaped by external stimulation.[131] Instead, the medical responses to alcohol abuse that took shape in the nineteenth century rested on the principles advanced by the new brain science, including the recognition of psychological interiority and the link between mental states and biological structures. Rush's deserved reputation as the father of the American temperance movement has obscured the fact that he resisted rather than promoted these intellectual trends. His later lectures betray deep contradictions, however, as he speculated about mysterious and irrational psychic phenomena—subjects of gothic speculation such as somnambulism, dreaming, phantasms, and the creative imagination—that seemed incompatible with his model of the republican man of sensibility. Even while resisting the weight of scientific inquiry, Rush's preoccupations nevertheless anticipated the direction of medical science.

In his lecture on somnambulism, for instance, at stake was his conviction that moral sensibility conformed to physical laws and that these laws could form the basis for shaping a perfect republican citizenry. The lecture included two newspaper clippings that expressed the romantic fascination of writers and poets with the power of dream states.[132] These newspaper clippings were gothic in the sense that the stories dwelled on the fundamentally mysterious nature of somnambulism, as sleepwalkers accomplished feats beyond their waking abilities. One clipping described a ten-year-old boy who had discovered that an owl had a nest at the top of the "old church steeple." Fascinated by the bird, the boy attempted to climb the building wall to see the nest, but failed. That night, however, "in the most profound sleep, he rose through the night, ascended the Gothic edifice, carried off the favorite object of his most earnest desire," and returned to his bedroom with the owl's nest.[133] Typical of romantic writing, the anecdote portrays somnambulism as a psychic phenomenon of intriguing potential, allowing a boy to enter a forbidding, dark interior and emerge with an owl's wisdom.[134]

In other lectures, however, Rush rejected more recent medical theories that portrayed the mind as creative and belittled the romantic fascination with the unconscious.[135] He taught that dreams, for instance, resulted from essentially mechanistic functions of the mind as it recycled waking ideas and sensations, a view predominant among Enlightenment thinkers in the mid-eighteenth century. He went so far as to argue that sleep was a "disease" and dreams were "morbid phenomena" that, consistent with his theory of the unity of diseases, were caused by an "inequality of excitement . . . in the blood vessels of the brain." In lectures on sleep, dreaming, somnambulism, and incubi, Rush was particularly concerned with immorality. Because the moral faculty is asleep while we are dreaming, for instance, "we shall dream of doing or saying things of an immoral nature, at which we should shudder in a state of complete and universal wakefulness." While "pious people are often much distressed at such dreams," Rush assured them that "there is no more immorality in them, than there is in striking a friend in the delirium of fever, or walking in our sleep." Nocturnal "seminal emissions," supernatural occurrences, seeing friends in "the most grotesque dresses," and other incoherencies of thought and feeling Rush explained by saying, "dreams may be considered as a low grade of delirium, and delirium as a high grade of dreaming."[136] For treating the diseases of sleep, Rush characteristically recommended bleeding.

Hallucinations were even more problematic. A common topic of theorizing in the new literature on mental illness and a common subject of popular gothic novels, Rush first gave a lecture titled "Phantasms" in 1809. Student notes on the "Phantasms" lecture suggest that Rush sought to dismiss, even ridicule, the subject.[137] For him, to accept that terrifying visions and delusions could erupt unsummoned and unruly into the mind's eye would be to accept that the moral faculty was subject to unpredictable forces. This idea violated the principle that was fundamental to Rush's vision of the place of medicine in a republican society: that moral development conformed to scientific laws and that republican citizens could therefore be shaped and molded. He attributed sightings of the supernatural to the mechanisms involved in visual and aural perception, and he reportedly said, "No more happens here than when pain is excited in the urethra from a stone in the bladder." His recommended therapy for phantasms, "bleeding, purging, low diet &c.," treated them as another form of fever.[138]

In his lecture on imagination, Rush similarly combined a mechanistic Lockean psychology and heavy republican morality with romantic psychic flourishes. He distinguished between "imagination" and "fancy." The imagination, he lectured, "ascends above the heavens, and explores the worlds that revolve around our earth, it descends into the regions of darkness, and beholds every form of moral depravity and misery—and all this it performs in the twinkling of an eye, thus encroaching . . . upon the omnipresence of the Deity." By comparison, the faculty of fancy was lesser and morally suspect. "Fancy occupies itself about phantasms or nonentities such as fairies and monsters, while imagination occupies itself exclusively about realities," he wrote.[139] But again, imbuing the mental faculties with creative possibilities raised conflicts with his views of the mind as entirely activated by sensation. In a late revision to his imagination lecture, the conflict is apparent in a single cryptic sentence that read, "The ideas that fill the imagination are derived chiefly from the eyes and the ears."[140]

Even if Rush doggedly asserted anachronistic explanations for altered mental states and the creative faculty of imagination, his theatrical readings of newspaper stories about gothic murders and somnambulists demonstrates that he was engaging topics of compelling popular interest. He initially developed these medical theories, after all, to present in lecture halls filled with several hundred young men. Most of these students came from towns much smaller than Philadelphia, then the nation's second largest city. Philadelphia offered Rush's students ready access not only to forward-looking medical texts, but also to literature, newspapers, magazines, theater, and other cultural attractions such as Charles Willson Peale's popular natural history and anatomy museum.

In Philadelphia theaters, for instance, medical speculation about intoxicated states of mind became a focus of popular interest. Just months after Rush's death in 1813, nitrous oxide debuted on the Philadelphia stage.[141] Discovered in 1799 by Sir Humphry Davy, its effects on the mental faculties enthralled contemporaries.[142] Initially, physicians and scientists ignored the potential anesthetic properties of the gas. What amazed Philadelphians in the 1810s was the "exhilarating" and fantastical feelings and perceptions it elicited in the mind.[143] Performances recreated Davy's famous experiments in which the doctor and his colleagues inhaled the gas and recorded their experiences. Theater shows illustrated how the mind-altering drug could elicit delusions that bordered on the fantastical. According to newspaper accounts, theatrical exhibits in Philadelphia consisted of a physician offering a brief lecture on the history and science of nitrous oxide followed by members of the audience inhaling the gas. Philadelphia's *Aurora General*

Advertiser described one 1820 exhibition as "highly interesting." Participants included, among others, a youth who burst into song, two gentlemen who began a vigorous fencing match with imaginary swords, and a man who "denounced the wrath of the Gods in very excellent Latin, worthy of the sybil, on the vices of the age." One young lawyer while under the influence of the gas "insisted that he possessed a logic so irresistible that he could cozen the devil himself." The gas so excited the mental faculties of the young man that he "concluded by assuring the auditors that if his satanic majesty were to send a message to him at that moment he would send his ambassador packing."[144] Intoxicating gas unlocked the mind's dark secrets.

Despite Rush's efforts to defend the sanctity of his psychology of republican virtue, the irrational, hidden realm of the psyche increasingly became an object of medical speculation, literary imagination, and popular entertainment. Rush's disease of the will diagnosis reflected these new romantic conceptions of the mind even as it sought to dismiss them. After 1810, he had come to describe the drunkard as being in the grips of an irrational compulsion, even resembling a somnambulist. "There was a Clergyman of this city when I was a student boy," Rush told his students, "who died from intemperance from the use of Ardent Spirits." When urged to stop drinking by his friends, the minister replied, "Were a keg of rum in one corner of a room, and were a cannon constantly discharging balls between me and it, I could not refrain from passing before that cannon, in order to get that rum."[145] Drunkards succumbed to a trancelike state in which their depraved desires defied all reason, a far cry from Rush's earlier mechanistic image of the moral thermometer.

If American physicians discarded his specific theories, it is also true that Rush's deep concern with the consequences of intemperance persisted as a defining characteristic of the still young American medical profession. Rush's disease of the will had offered a framework by which alcohol abuse might be treated as a mental disease, but in the years immediately following his death American physicians showed little inclination to try to cure drunkards of heavy drinking. They had moved on from his republican medicine, now a relic of an earlier time, and rejected the idea that the moral faculty could be an object of medical treatment. Instead, in the 1810s pathological drinking became an object of gothic speculation through the newly described disease delirium tremens. Rush would have dismissed the intellectual principles this new diagnosis rested on, but his lectures on moral derangement, somnambulism, phantasms, and imagination nevertheless anticipated physicians' intense fascination with it.

TWO

Discovering Delirium Tremens

"Of all the diseases to which the human race is subject," the physician Pliny Earle wrote in 1848, "there is none that more completely unmans its unfortunate victim." Writing for *American Journal of the Medical Sciences*, Earle described a disease with frightening physical symptoms that included vomiting, uncontrollable trembling, and seizures. But for Earle "these physical symptoms . . . are but little when compared with the mental phenomena resulting from them." Both the external senses and the faculties of the mind become deranged, as the patient is entirely lost to the "wayward, excited, and ungovernable imagination."[1] Earle depicted the hallucinations patients experienced, images "more varied in their forms and characters than are the designs of the artist, more diverse and unstable than the ever changing pictures of a phantasmagoria." The patient succumbs to horror as "animals of various kinds throng into his room, crouch before him, with threatening gestures, and grimaces the most frightful." He begins to believe frightening delusions as "enemies in human form spring up to bind, to drag to prison, to the tribunal of justice, to the rack, or to the place of execution, or perchance to shoot or to slay with the sword." Finally, Earle continued, the patient must confront "the phantoms of the ideal world, specters with gorgon heads, and bodies more hideous than those of the satyr or the fabled tenants of the lower regions, glower upon him with their eyes of fire, gnash their teeth in fiendish defiance, at length seize upon him, and he struggles with them in the full faith that he has encountered the devil incarnate."[2]

The affliction Earle pictured so graphically was first described by British physicians in 1813. It became a topic of intense interest to the American medical community soon after. In medical journals, doctors variously termed the condition mania a potu, mania a temulentia, delirium vigilans, delirium potatorum, brain fever, or the term most often used today, delirium tremens.[3] Although they disagreed on the best name for the disease, the symptoms were easily recognizable: uncontrollable trembling, violent seizures, growing paranoia, and, most distinctively, vivid hallucinations. In less than ten years, delirium tremens became a standard lecture topic at American medical schools, a common subject of published case histories and student dissertations, and a daily occurrence in hospitals. The attention doctors focused on this new disease represented a sharp break from previous medical practice. Before 1815, physicians had done very little to treat inebriates. Benjamin Rush's views on the importance of temperance to health and well-being were widely influential, but his writing had done little to transform the way doctors treated patients who were heavily intoxicated or suffering the severe consequences of heavy drinking. Only after his death did American physicians increasingly hospitalize inebriates. By the 1820s, delirium tremens dominated how American physicians conceived of, studied, and theorized about the more general problem of pathological drinking.

Why early nineteenth-century physicians suddenly devoted so much energy to delirium tremens is a challenging historical question. In one sense "delirium tremens" was simply a new name for alcohol-induced insanity, a relatively common occurrence that early American physicians had shown little inclination to treat. But in the 1810s the published case histories of delirium tremens differed so much from previous descriptions of alcoholic insanity that one twentieth-century historian argued that the condition was entirely new—that before 1815, drinkers did not experience these violent symptoms.[4] Early nineteenth-century doctors did not describe delirium tremens as new, however. They agreed that the common affliction had until then been "too much neglected by practical writers in medicine," as one physician put it, and so invisible in the medical literature.[5]

In part, the new diagnosis reflected advances in conceptions of human anatomy and disease that promised to revolutionize the practice of medicine. These advances derived especially from exciting theories developed in Paris, which eclipsed Edinburgh as the new center of Western medicine. Debates over the pathology of delirium tremens and its appropriate treatment rested on empirical evidence gathered during postmortem dissection, an emphasis grounded in the growing influence of physiology and pathological anatomy. Medical interest in delirium tremens peaked during the same years that regular dissection became an essential dimension of American medical education and doctors' professional identity. The doctors most interested in the disease tended to be young and eager to advance their careers within the burgeoning and competitive Philadelphia medical community. They spread information about the diagnosis through nationally and internationally circulating medical journals, a relatively new print genre that proliferated rapidly during these years. The disease thus both reflected and shaped profound transformations remaking American medicine.

A new way of seeing a common affliction, delirium tremens was as much a product of science as of culture. Although delirium tremens lay at the forefront of medicine, case histories lingered over sensational details irrelevant to the disease's pathology. Earle's wild description was typical of narratives that read like dark romantic tales peppered with specters, delusions, and horror. Among medical students and their professors, graphic stories of the hideous apparitions their patients witnessed betrayed a gothic fascination with a psychic landscape of depraved desires and supernatural beings: imaginative visions that Earle evocatively described as "the ever changing pictures of a phantasmagoria." Accounts portrayed delirium tremens both as a disease and as a sort of theatrical performance. Stories of intemperate patients struggling with "the phantoms of the ideal world, specters with gorgon heads, and bodies more hideous than those of the satyr" expressed physicians' conviction that pathological drinking "unmanned" unfortunate victims. Especially in the years surrounding the financial Panic of 1819, these dark romantic narratives evoked something of the concerns of young men crowding into Philadelphia with tenuous hopes that medical training would bring them social advancement and respectability. Drawing on the language and imagery of romanticism, these case histories thus expressed cultural concerns about the fragility of masculine achievement at a historical moment of tremendous uncertainty.

In the 1810s and 1820s, new medical beliefs and practices centered on delirium tremens can best be understood as a response to the imperatives, contradictions, and stresses of the emerging market economy. The delirium tremens diagnosis founded the medical conviction that habitual heavy drinking was a physical and pathological affliction, and it led to profound changes in how the medical profession responded to the ubiquitous problem of alcohol abuse. But physicians were drawn to this particular diagnosis because they saw in it a larger symbolic significance. In gothic case histories, physicians constructed a liberal morality tale of the male market actor struggling to contain the demons of his imagination, which had been empowered by depravity and indolence. Published in 1812, Benjamin Rush's widely ignored disease of the will theory had portrayed pathological drinking as evil, a derangement of the moral faculty and a dire threat to the virtue of the American electorate. By contrast, the literature on delirium tremens turned the drunkard's alcoholic demons into a metaphor for the struggle for success in an unstable economy. Treating the disease, physicians contemplated the common and frightening dangers of masculine failure. At least as importantly as any scientific advance, this concern with respectability, failure, and social downfall would shape the nature of medical responses to pathological drinking into the twentieth century.

An Alcoholic Disease

In the United States, physicians in the medical ward at the Philadelphia Almshouse initially led the way in describing, studying, and treating delirium tremens. Isaac Snowden, a resident physician in the almshouse, wrote the first American essay on the disease in 1814 and published it the next year.[6] Just two years later Joseph Klapp, a member of the Philadelphia Almshouse Board of Physicians, ignited controversy over the disease. In published essays, personal letters, and teaching sessions at the almshouse, he aggressively promoted a radical new cure involving harsh emetics.[7] Doctors administered the emetic orally with the patient restrained in a hospital bed (fig. 3). Klapp directed his students to keep the patient vomiting until a substance appeared that was "thick, ropy, and . . . about the consistence [*sic*] of boiled tar. Its colour is generally a light brown. Sometimes, indeed,

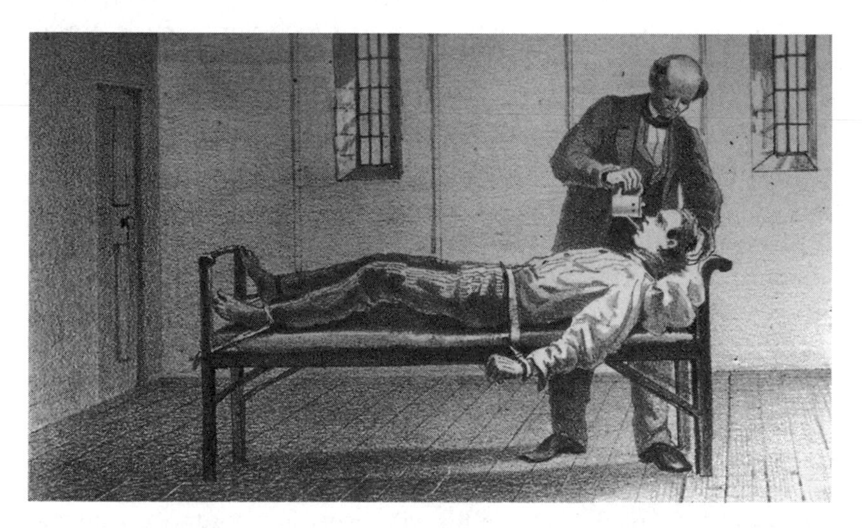

Figure 3. A midcentury drawing of a delirium tremens patient receiving treatment at the Pennsylvania Hospital. From Ebenezer Haskell, *The Trial of Ebenezer Haskell, in Lunacy, and His Acquittal before Judge Brewster in November, 1868* (Philadelphia: Ebenezer Haskell, 1869).

it exhibits an entire blackness."[8] Fatalities associated with the treatment prompted a string of medical journal articles debating its efficacy. In the 1820s, virtually every American essay written on delirium tremens voiced an opinion on "Klapp's cure."[9]

The years when delirium tremens became a subject of interest were a time of tremendous growth, excitement, and energy in the American medical profession, nowhere more so than in Philadelphia. Medical institutes, schools, and instructors multiplied, attracting hundreds of students each year. Influenced by new theories of pathology, modes of medical investigation, and professional imperatives, a new generation discarded old ideas and sought to remake the very meaning of medicine. Physicians in Philadelphia charted a new direction for the profession in a wave of medical journals, pamphlets, dissertations, and essays.[10] By the time Pliny Earle's essay appeared in 1848, physicians had produced an extensive literature on delirium tremens. American physicians wrote no fewer than ninety medical journal articles on the subject before the Civil War. Published in the 1880s, the sixteen-volume catalog of the Library of the Surgeon General's Office shows that the library held 441 American and European works on the subject of delirium tremens, including ninety-one books and dissertations published before 1865. By the 1880s the library counted at least 658 works on the topic of delirium tremens. That number does not include the many books and articles that contained discussions of the disease as part of larger topics, or the many unpublished medical dissertations produced each year at American medical schools.[11]

The delirium tremens diagnosis stood out as unique not just among mental illnesses, but among diseases in general. First, physicians based classification of the disease on its singular cause: habitual heavy drinking. In modern medicine diseases are commonly identified with their causes: "flu" is caused by a flu virus. But in the eighteenth century, diseases were primarily classified according to their predominant symptoms.[12] This understanding led physicians to lump alcoholic insanity with other forms of mental illness. In the mid-1790s, for instance, Erasmus Darwin included "delirium ebrietatis" in his elaborate hierarchy of delirium. He wrote, "The drunken delirium is nothing different from the delirium attending fevers except in its cause, as from alcohol, or other poisons."[13] For Darwin the cause was secondary to the outward symptoms of delirium. Likewise, in his *Observations on Madness* (1809), John Haslam included a number of case histories of lunatics whose condition he attributed to drinking hard liquor. Excessive use of spirits caused one man to sink into a derangement in which

> he conceived himself very nearly related to Anacreon, and possessed of the peculiar vein of that poet. He also fancied that he had discovered the longitude; and was very urgent for his liberation from the hospital that he might claim the reward, to which his discovery was entitled. At length he formed schemes to pay off the national debt: these, however, so much bewildered him that his disorder became more violent than ever.

The man died after four months' confinement.[14] For Haslam and Darwin these patients were mentally deranged, and the cause of that derangement—alcohol—was not especially significant in understanding their condition.

Rush blamed ardent spirits for one-third of the lunatics confined at the Pennsylvania Hospital. Despite his deep concern with intemperance, however, Rush, like all other American doctors, did not see alcoholic insanity as distinct from other manic disorders.[15] In his *Observations on Diseases of the Mind* (1812), he asserted that alcoholic insanity often manifested itself as a disease of the stomach, which he termed "derangement of the stomach." "Successive paroxysms of madness," Rush wrote, "occur most frequently in habitual drunkards; and they would probably occur much oftener, were they not prevented by a vicarious affection of the stomach, known by puking."[16] He classified and treated mania as a singular disease derived from multiple causes, however. In an 1811 dissertation, a student of Rush's at the University of Pennsylvania listed the principal causes of mania as "hereditary predisposition; abuse of spirituous liquors; violent and stimulating passions of the mind; abstruse study; unlimited exercise of the faculties; tumors compressing the brain." Mania consisted of a set of symptoms, but the disease might derive from any number of causes, of which alcohol was just one.[17]

Acknowledging that the disease had previously been "confounded with mania, or madness, from other causes," after 1813 physicians described delirium tremens as a distinct disorder and based this new distinction on its singular cause—excessive, habitual intoxication.[18] Writers occasionally argued that all forms of intoxication from "narcotic stimulation," including opium use, could cause the disorder, but alcohol remained the overwhelming focus of medical concern in Philadelphia throughout the antebellum period. Delirium tremens differed from the many other mental and physical diseases resulting from heavy drinking in that, as the British doctor Robert Macnish put it, "it originates *solely* in the excessive use of stimulating liquors, and is cured in a manner peculiar to itself."[19] Delirium tremens thus stood out in an era when mania might originate in fever, religious enthusiasm, the moon, or excessive concentration. In 1827 one Philadelphia physician commented, "The remote causes of delirium tremens are capable of being defined with more distinctness than . . . any other disease to which humanity is subjected."[20]

Identifying delirium tremens with a single cause had a number of consequences, including the creation of a new category of patient. Hospital doctors often referred to the remarkable violence of the affliction and the unique challenges these patients posed. The disease had the potential to wreak havoc in a hospital ward. Commonly, case histories described patients admitted with various complaints—pneumonia, a sore leg, or cholera, for instance—who after a day or two of recovery developed violent delirium tremens. At the Pennsylvania Hospital, Benjamin H. Coates reported that particular care had to be taken with fracture patients who were heavy drinkers. Such a patient "frequently escapes from his bed, and endeavours to walk seeming altogether insensible of pain in a limb which is frequently bent or nearly at right angles and incessantly agitated, at the place of fracture," wrote Coates. These scenes were so common that physicians briefly theorized that broken legs were an "exciting cause" of delirium tremens. Coates thus recommended that hospital administrators determine the drinking habits of all patients on admission. Habitual drunkards could then be sedated with laudanum or strapped into bed.[21] He acknowledged this determination might be challenging, given social norms: "My own experience would lead me to believe that the fear of shame will very generally prevent an acknowledgment by the patient of the whole amount of the liquor to which he is accustomed."[22] Patients also knew that the Pennsylvania Hospital treated only the "worthy" poor and may have feared that if they admitted how much they drank they would be expelled. The hospital did make Coates's proposed change. In the 1820s, admittance forms inserted a new column noting patients' drinking habits as "temperate" or "intemperate." Treating delirium tremens in an institutional setting thus led physicians to create new categories and shape new practices around the unique challenges of inebriates.

Alcoholic Apparitions

The delirium tremens diagnosis was also distinctive in the way physicians chose to describe the symptoms of alcoholic insanity. Nineteenth-century physicians understood delirium tremens as moving through three basic stages. Acute insomnia, nausea, vomiting, and constipation commonly appeared in the first stage, but uncontrollable trembling especially marked the onset of the disease: "there is always . . . nervous irritability, watchfulness and trembling of the whole body, but more particularly of the head and hands."[23] Doctors dwelled on patients' wild facial expressions: "The eyes are red and furious, never fixed, but incessantly wandering from object to object."[24] Soon patients were overcome with paranoia: "The mind becomes deranged, more particularly through the night. . . . when in bed appears restless, and in a short time, unless confined by force, will get up and put on his clothes, walk the room, manifest great anxiety about his affairs, or the safety of his person."[25] These symptoms built into a second stage characterized by a violent delirium lasting several days to a month. Typically, published case histories described the symptoms in great detail:

> In some instances they imagine they see some disgusting and loathsome animal in the room; as rats, mice, or snakes, which they suppose are come to do injury to their persons or property. Occasionally they imagine they see some frightful object, as the devil, who, they suppose, has come to take them; which occasions almost insupportable fright, manifested by a violent trembling of the whole system, expression of fear and horror in the countenance, and anxious cries for help. Sometimes they fancy that they hear remarkable noises in the room, or at a distance; and occasionally they alternately sing, pray, and rehearse passages of scripture.[26]

Philadelphia physicians often said that the sine qua non of treatment was to get the patient to sleep.[27] Most patients did eventually collapse into slumber, waking up in control of their faculties. Acute cases progressed to a violent third stage of epileptic seizures and a crescendo of insanity. The onset of seizures often led quickly to death.

Descriptions of their patients' visual hallucinations represented a sharp departure from the past. Rush never associated hallucinations with his "derangement of the stomach," for instance.[28] In delirium tremens case histories, accounts varied widely from brief mentions of phantoms or vermin to extended "reveries" in which patients participated in involved and intricate delusions. Writing in 1819 in support of Klapp's emetic cure, the physician Daniel Drake elaborated at great length on the case of a man who believed he was on an American naval vessel under attack by a French privateer.[29] Some accounts described patients who seemed eerily possessed or who behaved like somnambulists. A shoemaker, for instance, sat in a dark room for days going through the motions of making shoes and talking to absent people.[30] More often, doctors described terror and paranoia. Typically the hallucination involved some animal or supernatural creature representing a mortal threat. Rats, mice, snakes, birds, wild beasts, armed soldiers, evil men, devils, demons, and in one case a cow standing on its hind legs pursued these unfortunate patients.

These tales of alcoholic nightmares appear throughout the surviving records of the Philadelphia medical community, including student lecture notes taken at the University of Pennsylvania. Professor Nathaniel Chapman included fanciful descriptions in his lectures in the 1820s, and unpublished student dissertations describe supernatural creatures and violent insanity.[31] Written in 1824, one student's description typified the narratives that circulated at the medical school:

> Frequently these imaginations are filled with objects of dread and horror, [such] as monsters or evil spirits, whose intentions they suppose are to destroy or carry them off to a place of torment. Again, they fancy they hear strange noises in some corner of their own, or in an adjoining room, [such] as the sound of dying persons. Or that they see spots of various colours, or balls of fire floating through the atmosphere.[32]

A medical student wrote in 1821 that in comparison with other mental afflictions, mania a potu stood alone: "None is more formidable in its nature, or in its consequences more terrible."[33]

Why did doctors devote so much attention to patients' alcoholic hallucinations? One answer is that hallucinations were a common subject of speculation in the most influential European treatises on mental illness.[34] These authors provided new theories on why and how apparitions could derive from various physical and mental afflictions. In his *Treatise on Insanity* (1806), for instance, Philippe Pinel wrote that a diseased imagination could produce "fantastic illusions and ideal transformations," and he listed patients haunted by devils, demons, angels, ghosts, and phantoms.[35] In *Inquiry into the Nature and Origin of Mental Derangement* (1798), Alexander Crichton distinguished between voluntary acts of the imagination, which occur while reading literature, for instance, and involuntary acts, which include images produced in dreams, the delirium of fevers, insanity, and religious fervor. When delirium was beginning to set in, Crichton noted, patients who shut their eyes or were placed in the dark would see "a crowd of horrid faces, and monsters of various shapes, grinning at them, or darting forward at them."[36]

These new theories transformed hallucinations into a symptom. Philadelphia doctors commonly attributed the horrible phantoms associated with delirium tremens to a diseased "fancy" or "imagination." One physician reported of a typical patient, "visions were constantly floating before his diseased imagination representing death, under every shape, with complicated horrors."[37] These observations were very similar to those of European writers. Crichton argued, for instance, that apparitions are images of the imagination that overwhelm the senses from within:

> The belief in the reality of the phantoms of the imagination arise[s] either in consequence of causes which prevent the impressions of external objects from reaching the brain with a due degree of force, or else from the images of the imagination having acquired such a degree of force from frequent repetitions, as to be superior in their effect to those derived *ab externo*.[38]

This effect, he went on to argue, is why phantoms often first appear in the dark, because then the mind is subject only to the phantoms of the imagination, without the impressions gathered from the outside world to counterbalance them. Philadelphia physicians' interest in hallucinations had lagged behind that of their European counterparts. Crichton's influential work was republished in Philadelphia in 1798, but Rush, the leading American expert on mental illness, introduced "phantasms" into his course of lectures only in 1809, and even then he characterized the phenomenon as trivial.[39] The intense interest in alcoholic hallucinations only a few years after Rush's death in 1813 illustrates that Philadelphia physicians embraced the new dark and romantic view of the human mind being advanced by European authors.

This shift is evident even in internal hospital records. Between 1804 and 1828, at the Pennsylvania Hospital, where Rush worked for thirty years, physicians preserved descriptions of significant case histories in a manuscript volume. In the first decade of the nineteenth century, several cases described alcoholic insanity without any mention of hallucinations. In one, a fifty-eight-year-old seaman of "intemperate habits in drinking" was admitted for a badly broken leg. The patient appeared headed for recovery, but on the morning of the fourth day he "was found standing on his sound leg and supporting himself by the bedstead, having removed in delirium which came on in the night, the splints and dressing entirely from his thigh." One 1808 case history attributed a case of delirium in an inveterate drunkard to "mania from the abstraction of stimuli." Hospital physicians treated the patient with opium and alcohol, which came to be prescribed for delirium tremens in later years.[40] In the 1820s, by contrast, case histories identified delirium tremens by noting behaviors like "picks [at] the bed clothes and his nose, nervous and paces, talks much, imagines sees persons and things which are not present," and "by night he had become very noisy and was much disturbed by phantoms."[41] The presence of phantoms became the most characteristic symptom of alcoholic insanity.

If hallucinations became a symptom, however, physicians' descriptions of them were anything but dry and clinical. Many read like ghost stories, and the sensational details doctors related were essentially voyeuristic. Credited with publishing the first description of the disorder, the British physician Samuel Pearson described how phantoms relentlessly haunted one patient, accusing him of committing a murder twenty years before. The man regained his sanity only after traveling fifteen miles to the grave of the victim.[42] In 1815 Snowden described the typical delirium tremens patient: "He intreats not to be left alone, points to men and devils armed with daggers and other weapons, whom he expects every moment to see commencing his destruction. He is in horror of ten thousand evils, and will endure no contradiction, no refusal of compliance with his demands!"[43] Physicians sometimes acknowledged that their blow-by-blow descriptions of the hallucinations were medically irrelevant. Justifying their lengthy narratives, they repeatedly referred to delirium tremens as "peculiar," "interesting," and even amusing. As one doctor wrote, "Among the varieties of mania which I have observed, none have arrested my attention, and interested my feelings so much as mania a potu."[44] Drake interrupted his case history to describe "a paroxysm of reverie, so interesting in its character, that . . . I flatter myself you will be amused with its history."[45]

Doctors often devoted as much time to relating amusing and theatrical stories as to offering methods of treatment or theories of pathology. This indulgence in gothic imagery is strongly evident in Joseph Klapp's 1817 essay, which touched off so much controversy. He devoted nearly a quarter of the twelve-page article to detailing the strange delusions of one patient. Mr. G., a forty-year-old carpenter, had been admitted to the almshouse for "mental derangement . . . occasioned by the excessive use of ardent spirits."[46] During his derangement, two imaginary evil men harassed Mr. G., persistently pulling at his bedclothes with long iron hooks. After chasing them away, Mr. G. returned to his bed, only to be tormented by band music. Looking out on the street, Mr. G. was then struck by the vision of a splendid, ornamented edifice. As he wandered outside to gaze at the beautiful domed structure, the two men who had originally been harassing him with hooks invited him to perform with them in a farcical theatrical production. When Mr. G. refused, they began beating him violently.[47]

Why relate these intimate details? Why were Philadelphia physicians inspired to offer such long stories? Pinel and Crichton did not include similar descriptions. Philadelphia physicians' writings demonstrate that popular fascination with spectral illusions was an important dimension of the medical interest in delirium tremens. In the first two decades of the nineteenth century, ghosts, phantoms, and delusions appear throughout Philadelphia's literary and theatrical culture. Apparitions were common in popular British gothic novels, for instance, and in the work of Philadelphia author Charles Brockden Brown.[48] Gothic melodramas, such as *The Mysterious Monk* and *The Castle Spectre*, were enormously popular in Philadelphia theaters during the 1810s.[49]

Philadelphia's preoccupation with phantoms illustrates the porous boundary between medical inquiry and popular culture. One measure of the popular fascination with hallucinations was a new form of theatrical entertainment called the phantasmagoria, which debuted in Philadelphia in 1809, just five years before delirium tremens first appeared in the city. First presented in Paris in 1798, the phantasmagoria featured an improved version of the magic lantern that magicians and entertainers had long used to project fanciful images, and often for supernatural effects.[50] Invented by Étienne-Gaspard Robertson, a Belgian engineer who trained at Edinburgh, the new lantern projected a stronger, more concentrated beam of light. Robertson also used a movable translucent screen and put the lantern on rollers, permitting the illusion that images were moving through the air and growing larger and smaller. His phantasmagoria shows used these techniques to project images of ghosts, devils, and other supernatural objects of horror.[51] By 1802 the shows had become wildly popular in London.[52]

More than using new projection technology, the phantasmagoria turned contemporary medical and literary exploration of the psyche into a stage show. Billed as "experiments in natural philosophy," shows commonly began with a lecture drawing on the recent medical literature that described specters as hallucinations.[53] By highlighting the fallibility of human perception, they popularized one of the central assertions of the new brain science—that apparitions were products of diseased mental faculties or senses. But from its inception, the phantasmagoria always reveled in the gothic cult of mystery. In his memoirs, Robertson described the evening entertainment as it was presented in Paris.

> The doors were locked with a resounding crash. The walls of the hall were lined with the skulls and bones of departed monks. The sudden extinction of the light, along with the sense of imprisonment, plunged the spectators into the most profound gloom, as if they were already in the tomb, among the shades. The eyrie tones of the harmonium then accompanied the sudden

> appearance of ghosts and specters, tenants of the grave and of hell itself, fluttering promiscuously and grimacing horribly among the spectators.[54]

Similarly in London, the Scottish scientist Sir David Brewster reported that the most impressive part of the phantasmagoria shows was the exhibition of "specters, skeletons, and terrific figures which . . . suddenly advanced upon the spectators, becoming larger as they approached them . . . the spectators were not only surprised but agitated."[55]

In the 1810s, the phantasmagoria became a staple of theatrical entertainment in Philadelphia. Advertisements described these shows as republican exercises in fostering a more enlightened citizenry. In 1809, Rubens Peale was the first to present the phantasmagoria in the city. The son of the painter and proprietor of the Philadelphia Museum, Charles Willson Peale, Rubens declared that the presentation would "enlighten and guard people against certain superstitious ideas they may have imbibed respecting witches and wizzrds [*sic*], which, in past ages, have kept the human mind in fetters."[56] The "magic lanthorn" soon became a regular attraction at Peale's museum as part of Rubens's evening presentations of various "philosophical" amusements.[57] The popularity of the phantasmagoria grew in the nineteenth century as magic lanterns appeared in popular museums, magic shows, and other theatrical productions. The phantasmagoria popularized new medical theories of the psyche, and physicians in turn sometimes used the magic lantern as an analogy in describing the mental experience of hallucination, such as when Earle wrote that delirium tremens hallucinations are "more diverse and unstable than the ever changing pictures of a phantasmagoria."[58] The literature on delirium tremens betrayed the same fascination with spectral illusions evident in this new scientific theater.

This fascination also found expression in books and essays on the science of apparitions written for a popular audience, and some of these works contained descriptions of hallucinations caused by delirium tremens. Written mostly by physicians, two of the most influential were Samuel Hibbert's extensive treatise *Sketches of the Philosophy of Apparitions, or An Attempt to Trace Such Illusions to Their Physical Causes* (1824) and James Alderson's "On Apparitions" (1810). Alderson's essay appeared in the *Edinburgh Medical and Surgical Journal*, the same journal in which the first essays describing delirium tremens appeared in 1813.[59] Both Hibbert's book and Alderson's article circulated in Philadelphia, and physicians regularly checked them out of the medical library of the Pennsylvania Hospital.[60] In medical journals, physicians writing on delirium tremens cited both these works as offering illustrative case histories. Likewise, Hibbert often cited the new disease delirium tremens in his explanations of ghost sightings. The delirium tremens disease model was thus always integral to, as Hibbert put it, the "general interest excited on the subject of apparitions."[61]

As with the phantasmagoria, Hibbert's book popularized some of the key assumptions of the new brain science. He begins by advancing the Enlightenment premise that "apparitions are . . . nothing more than ideas or the recollected images of the mind, which have been rendered more vivid than actual impressions."[62] He addressed many of the physical and mental afflictions that were thought to cause ghost sightings, drawing on the most influential European writers of the day, including Pinel and Crichton, as well as the work of physicians John Ferriar and John Haslam. Most of the book, however, consisted of detailed and fanciful accounts of demons, fairies, elves, spirits of the departed, and all manner of supernatural beings. Interestingly, Hibbert's frequent protestations that he sought only scientific explanations for them heightened the allure of his descriptions.

Grounded in recent medical theory, this elite and popular preoccupation with hallucinations caused readers to see very old texts in new ways. On Philadelphia stages in the 1810s and early 1820s, for instance, theaters commonly paired productions of William Shakespeare's *Hamlet* with a new farce called *The Ghost, or The Affrighted Farmer*. The rationale for this particular pairing presumably was the popular appeal of specters.[63] Although grounded in the new brain science, Hibbert's book included a long discussion of the ghost in *Hamlet* and relied heavily on another early seventeenth-century text, Robert Burton's *Anatomy of Melancholy*. Originally published in 1621, Burton's *Anatomy* had been out of print for over a century. Interest in the text was revived in large part through the efforts of the British physician John Ferriar, who himself published a theory of apparitions in the early nineteenth century. Reprinted in 1800, Burton's *Anatomy* was cited by many romantic-era writers and has been continuously in print ever since.[64] The book also circulated widely in Philadelphia in the 1810s and 1820s.[65] Hibbert considered Burton as central in making the connection between apparitions and diseased states of mind, even though those connections were based on antiquated theories of medicine.[66] Burton cited the "corrupt imagination" when explaining the hallucinations and delusions of maniacs. Quoted by Hibbert, Burton described individuals suffering one common type of mania as

> more than ordinary suspicious, more fearful, and have long, sore, and most corrupt imaginations; cold and black, bashful, and so solitary, that they will

> endure no company. They dream of graves, still and dead men, and think themselves bewitched or dead. If the symptoms be extreme, they think they hear hideous noises, see and talk with black men, and converse familiarly with devils, and such strange chimeras and visions, or that they are possessed by them, and that somebody talks to them, or within them.[67]

Burton's seventeenth-century description bore a striking resemblance to the sensational accounts of delirium tremens circulating in Philadelphia in the 1810s and 1820s.

The consequence of all of this popular and scientific speculation was to popularize the idea that phantasms emanated from within the mind, and that these visualizations derived from organic causes, like disease or injury. Hibbert's fanciful writing made the difficult ideas of Pinel and Crichton more accessible to readers. Certainly, Crichton and other writers who wrote about the nature of hallucinations would dismiss superstitious beliefs in witchcraft, ghosts, and other specters of ignorance, but as literary scholar Terry Castle puts it, "Once an apparition-producing faculty was introduced into the human psyche, the psyche became (potentially) a world of apparitions." For Castle this move had the paradoxical effect of making ghosts even more real, since the effect of placing the supernatural into the realm of psychology was to "demonize the world of thought."[68] The new science of apparitions posited a human mind that contained frightening mysteries, always threatening to erupt out of dark shadows.

Delirium tremens introduced a compelling moral dimension to these psychic demons. As with forms of insanity, so too with hallucinations, before the delirium tremens diagnosis the actual cause of a patient's hallucinations was not particularly significant. But delirium tremens created a new category of frightening demons derived specifically from habitual heavy drinking. In this distinction, delirium tremens was also related to the romantic literature addressing narcotic intoxication. Thomas De Quincey's *Confessions of an English Opium-Eater* (1821), was widely read and cited in nineteenth-century Philadelphia medical journals and newspapers, and it spurred doctors' interest in opium dreams.[69] An 1822 review published in a Philadelphia medical journal cited the author's dreams as the principal subject of the book and the "main part of the author's suffering." The passages excerpted from De Quincey, particularly elaborate and fanciful, resembled the fantastical reveries published in delirium tremens literature:

> I was stared at, grinned at, chattered at, by monkeys, by paroquets, by cockatoos. I ran into pagodas, and was fixed for centuries, at the summit, or in

> secret rooms; I was the idol; I was the priest; I was worshipped; I was sacrificed. I fled from the wrath of Brama through all the forests of Asia: Vishnu hated me: Seeva laid wait for me. . . . I was buried, for a thousand years, in stone coffins, with mummies and sphinxes, in narrow chambers at the heart of eternal pyramids. I was kissed, with cancerous kisses, by crocodiles; and laid, confounded with all unutterable slimy things, amongst reeds and Nilotic mud.[70]

The review finishes with an analysis of the physiological roots of the horrid dreams and suffering the opium eater endured. These explanations also mirrored the physiological explanations for the sightings of apparitions that doctors detailed in delirium tremens literature. In the 1810s, delirium tremens did not yet have poets. No ecstatic visionaries drew on alcoholic insanity to glimpse the secrets of the universe or the mysteries of the East. But delirium tremens was also far, far more common than opium or nitrous oxide intoxication. While alcohol was ubiquitous in American life, opium was comparatively scarce and nitrous oxide extremely rare. Cases of opium intoxication were uncommon in medical institutions. By contrast, delirium tremens quickly became a daily occurrence, inviting speculation and controversy. In shaping medical and popular conceptions of alcohol and drug use, the alcoholic demons that emerged from medical literature would prove far more sinister and influential than opium dreams or phantasmagoric specters because they were far more prevalent.

The medical theories and practices that transformed inebriate care in the nineteenth century thus derived in part from this romantic fascination with narcotic intoxication, insanity, and imaginative phenomena. Alcoholic insanity had been so commonplace that it was all but invisible to physicians. But the delirium tremens disease model outlined by British physicians in 1813 connected a previously banal condition to these diverse cultural forms of popular romanticism that reveled in dark gothic explorations of psychic wonders. Philadelphia physicians' interest in the disease came in part out of this enthusiasm for the stories of imaginative hallucinations, maniacal ravings, and elaborate reveries that they detailed in case histories. In popular culture and common parlance, these alcoholic demons persisted as one of the most enduring legacies of the delirium tremens diagnosis.[71]

The Anatomy of Intemperance

In connecting hallucinations to organic causes, the rise of the delirium tremens diagnosis also illustrates a more fundamental shift in American medicine. When Snowden wrote his 1815 article on mania a potu, the Philadelphia medical community was on the verge of what one historian has termed a "mania for anatomy."[72] This development would have profound consequences for medical theories related to pathological drinking. As doctors increasingly turned to postmortem dissection to explore the nature of disease in general and delirium tremens in particular, they often referred to "intemperance" to explain inflammation of the internal organs. It was through this scientific endeavor that physicians began to theorize that habitual heavy drinking could be a physical disease.

In the early nineteenth century, a rapidly expanding transatlantic medical community organized and disseminated knowledge production in a spate of new medical journals. In British journals such as the *Edinburgh Medical and Surgical Journal* and in new American publications such as the *American Medical Recorder*, the *Eclectic Repertory and Analytical Review*, and the *Philadelphia Medical and Surgical Journal*, elite doctors shared clinical observations, illustrative case histories, and postmortem dissections. They declared their allegiance to a practical and "fact-based" mode of medical practice, which they contrasted with the discredited grand medical theories of the eighteenth century. Samuel Pearson began his 1813 essay on delirium tremens by saying,

> Multifarious and repugnant theories on the science of life still continue to agitate the medical world. . . . A medical review will convince anyone how the faculty worry each other . . . about their different dogmas, with much injury to themselves and patients. For the above reasons I disavow all theory, and briefly state the circumstances as they occurred to me at the patient's bed-side.[73]

Snowden introduced his essay with a similar declaration, promising to reject "all specious display of hypothesis" and strictly adhere "throughout the communication, to fact."[74] Medical "fact" relating to disease increasingly meant visual evidence. One of the most arresting facts associated with delirium tremens were the patients' wild ravings, so much so that hallucinations became known as the most recognizable and distinctive symptom of the disease. In his influential 1827 article Coates wrote, "The delirium . . . is of so peculiar a kind, as of itself to furnish almost universally the means of distinguishing the complaint. Besides tremor and watchfulness, it is marked by the occurrence . . . of *apparitions*."[75] One of the justifications physicians gave for describing hallucinations in great detail was that they were simply relating facts.

Anatomical investigation in Philadelphia began when physician William Shippen built the city's first anatomical theater in the 1760s. At the University of Pennsylvania, Caspar Wistar held the professorship of anatomy from 1790 to 1810. At the same time that Rush published the first book-length American work on psychiatry, Wistar wrote the first American textbook on anatomy, the two-volume *System of Anatomy* (1811, 1814). By the 1820s, anatomical investigation was at the center of medical science in Philadelphia. Private anatomy instructors and schools proliferated, such as the Philadelphia Anatomical Rooms, supplementing the anatomy theaters at the Pennsylvania Hospital and the University of Pennsylvania.[76]

Medical research and education increasingly centered on studying morbid anatomy to identify the cause, or "seat," of various diseases.[77] Because physicians saw liquor as a highly stimulating substance, habitual drinking offered a ready explanation for any inflammation in the internal organs. Anatomists acknowledged that dissection did not grant insight into all diseases. The influential British anatomist Matthew Baillie wrote in 1793, "There are some diseases which consist only in morbid actions, but which do not produce any change in the structure of parts: these do not admit of anatomical inquiry after death."[78] But dissection did offer clear insight into the physiological consequences of intemperance. The findings of eighteenth-century anatomists in relation to alcohol had been dramatic. Citing famous European physicians, Anthony Benezet wrote in 1774 that ardent spirits "parch up and contract the stomach to half its natural size, like burnt leather, and rot entrails, as is evident . . . by opening the bodies of those persons who are killed by drinking them."[79] While these findings were widely accepted in the eighteenth century, only a few elite anatomists reported them. Physicians, most notably Benjamin Rush, used these secondhand reports only loosely and eclectically as evidence of liquor's impact in various diseases.

The literature on delirium tremens dramatically expanded the speculation and investigation of the morbid anatomy of the inebriate. After 1817, Klapp's assertion that delirium tremens derived from a morbid inflammation of the stomach became the prevailing view in Philadelphia for the next decade. Echoing though not citing Rush's description of derangement of the stomach, Klapp argued that delirium tremens resulted from repeated alcoholic stimulation of that organ, evinced by the fact that inebriates were prone to puking. In 1821 James Martin Staughton carried this observation into the dissection room, testifying in his influential dissertation that the "viscus in inebriates is generally affected, and in *post mortem* examinations is found sometimes inflamed—sometimes with thickened coats—sometimes in a scirrhous state, and sometimes contracted into a small pouch."[80] Students at the University of Pennsylvania commonly noted inflammation of the stomach in the postmortem dissections of delirium tremens victims.[81]

Klapp's focus on the stomach derived in part from the particular social concerns that had run strong in the Philadelphia medical community since Rush. In the late 1810s, delirium tremens was only one of a number of stomach diseases that drew interest from Philadelphia physicians.[82] Expressing alarm, they associated these diseases with excesses of diet. "Mankind ever anxious in the pursuit of pleasure," wrote one medical student, "have reluctantly admitted into the catalogue of their diseases, those evils which were the immediate offspring of their luxuries."[83] In addition to the many dissertations on delirium tremens, medical students at the University of Pennsylvania produced a spate of essays on dyspepsia and hypochondria—both considered gastric diseases associated with hallucinations.[84] "In these times of almost Persian luxury and effeminacy," wrote one student, "to appease the perverted craving of that idol of the sensualists worship, the stomach, it is not to be wondered at, that we should pay the penalty of one having thus transgressed the limits of Nature simply by incurring a multitude of diseases unknown to our more temperate ancestors."[85] These students expressed a republican disdain for sensuality that Rush would have been proud of.

The stomach became a primary concern of medical practice. Philadelphia medical journals promoted gastric treatments for insanity as well as delirium tremens. The authors argued that morbid anatomy demonstrated "that intestinal irritation, or organic injury of their internal coat, is capable of inducing every grade of cephalic and mental affection, from the slightest headach [*sic*] to the wildest ravings of mania, or the most sullen torpor of idiocy."[86] The earliest writing on delirium tremens, in the 1810s, commonly noted diet. One physician reported of a patient, "I was told that for several weeks previous to his disease, he had had a very great thirst, and voracious appetite, particularly at night; that on going to bed, he generally took with him, a quart of water, a half a loaf of bread, and a half pint of gin."[87]

In the 1820s, this preoccupation with effects of alcohol on the stomach got a significant boost from the growing influence of French medical theorists, most notably Xavier Bichat and François Broussais.[88] French physiology formed the primary intellectual framework by which physicians conducted postmortem dissection. These theories described the natural laws that governed the body. Physiologists believed that only extensive study of anatomy and experience with dissection could provide the requisite knowledge of these inner workings. "The human system is a complicated animal mechanism," Samuel Jackson explained. "It consists of numerous organs, each having peculiar functions; these organs are arranged into apparatus for the performance of certain offices; each organ is itself a compound, resolvable into elements called tissues or membranes, each tissue having a peculiar nature, actions and functions."[89] In a fully physiological approach to medicine, nothing could replace the practical knowledge that dissection provided.[90]

Prominent physicians in Philadelphia began to identify themselves closely with the new French medicine, and among them was the city's leading doctor, Nathaniel Chapman, professor at the University of Pennsylvania. The most ambitious and wealthiest American medical students traveled to Paris.[91] In the early 1820s a small group of young and well-connected physicians used their fluency in French medical theory to forge elite careers. These doctors included John Godman, who wrote several influential works on pathological anatomy before dying unexpectedly in 1830; William E. Horner, an adjunct professor of anatomy at the University of Pennsylvania who would go on to become a full professor and dean of the medical school; Benjamin H. Coates, who won appointment as a principal physician at the Pennsylvania Hospital in 1828 and edited the *North American Medical and Surgical Journal*; Samuel Jackson, a physician at the Philadelphia Almshouse Infirmary who in the 1830s would become professor of the Institutes of Medicine at the University of Pennsylvania; and William Wood Gerhard, a principal physician and influential teacher at the Philadelphia Hospital (formerly the Philadelphia Almshouse Hospital) in the 1830s.[92]

More than any other single physician, Horner dominated anatomical instruction in Philadelphia.[93] One of Rush's last students, Horner had an interest in the relation between habitual drinking and chronic stimulation that went far beyond Broussais or any of the other French anatomists.[94] In his *Treatise on Pathological Anatomy* (1829), the first book-length American work on pathological anatomy, Horner wrote, "I may now state that M. Broussais, in his History of Chronic Inflammations, has, in my opinion, rendered a service to pathological medicine. . . . That chronic inflammations of organs essential to life, destroy more individuals than pestilence and the sword added together."[95] Broussais had observed that the chronic inflammation caused by habitual drinking destroyed the lining of the stomach, but Horner expanded considerably on this observation: "I have found the mucous coat thickened and dense, without any remarkable contraction of the stomach . . . so reddened by numerous capillary vessels injected with blood, that at the distance of a few feet they appeared, when the distinction of the individual capillaries was lost in the distance, like red streaks."[96] Horner even included color plates to illustrate the gradations of inflammation in the tissue of the stomach as it became progressively diseased by the habitual use of ardent spirits.[97]

As illustrated by Horner's theories, and despite emphatic rejections of "multifarious and repugnant theories," many of the central concepts taught in Philadelphia in the 1820s were not a complete departure from the medical theory that Rush had developed. Rush had studied with the great London anatomist William Hunter, and he strongly advised young medical students "to open all the dead bodies you can, without doing violence to the feelings of your patients."[98] But in his Institutes lectures, Rush cautioned that the science had limitations, saying, "Anatomy may be compared to the outside of a picture."[99] Rush believed that the infusion of life into the physical structures of the body represented the most important subject of physicians. To illustrate the inside of the "picture," he resorted to long and labored analogies, biblical and literary references, and abstract models of the psyche unconnected with physical structures. For Rush the faculty of reason, not the anatomist's scalpel, was the best vehicle to advance medical knowledge.[100]

When Rush died in 1813, critics rejected his persistent use of analogies over the hard evidence provided by dissection.[101] Yet, like Rush, the physiologists presented a view of the individual as formed by impressions from the outside world according to the principle of sensibility. The complex apparatus of the body was further bound together in a system of sympathy and sensibility: "The organs and apparatus are connected in two ways. The one depending on the function from the communication of impressions and actions made on one to another, which is through sympathy."[102] In contrast to Rush, the new French theorists presented a more organic and biological medicine. Xavier Bichat was particularly influential in identifying the various types of tissues that made up the body and describing the gradations of inflammation in them. Philadelphia's physiologists argued that all disease derived from morbid stimulation. "All impressions made on the organs must be received by its tissues," wrote Jackson. "Every impression produces some change or modification in the action of the tissue to which it is applied; and every change of action must be accompanied by change of structure." The implication of these theories was that highly stimulating substances, especially ardent spirits, were particularly harmful to human health. Jackson continued, "Disease then, the result of abnormal impressions, is a change in the action and structure of the tissue of an organ."[103]

The doctrines of the French physiologists were heatedly debated in the

1830s, but interest in the influence of liquor on the body persisted long after Broussais, in particular, fell out of fashion.[104] Published autopsies often featured anatomical observations of the damaging effects of liquor on the internal organs. Physician Samuel Annan regularly contributed case histories from the Baltimore Almshouse Hospital to Philadelphia medical journals. In 1839 he described a postmortem dissection of an "old sailor of intemperate habits" who died following a bout of dysentery after seven years in the almshouse. Annan noted extensive gastric irritation and a diseased state of the brain. He observed,

> A large majority of those who die in this institution have been long addicted to habits of intemperance; and in every instance more or less of the effects of chronic gastritis is discovered. . . . [T]he shades of colour which principally belong to this disease, are the gray slate colour, the brown colour, and the more or less deep black colour. These discolourations, with preternatural redness, and thickening and softening of the mucous membrane, are the common morbid appearances.[105]

In the case of a drunkard who died of apoplexy and palsy, Annan's autopsy demonstrated "a well marked example of the effects of habitual drunkenness upon the brain."[106] Physicians thus reached a well-developed medical consensus on the "morbid appearances" of internal organs diseased by drink.

In his 1827 article "Observations on Delirium Tremens," Benjamin H. Coates stated most forcefully the implications of the new French medicine for the medical investigation of intemperance. Coates synthesized the growing medical literature and drew on his own extensive experience treating the disease at the Pennsylvania Hospital to urge physicians to abandon Klapp's cure, which he believed to be poisonous.[107] He also developed a detailed discussion of the pathology of delirium tremens. He argued that "the known operation of spirituous liquors, and of opium, on the great organs of the human body, consists in an excitement of the circulation, accompanied by a depression of the cerebral functions." Repeatedly applied to the system, alcohol habituates the body to narcotic stimulation: "The well known *accommodating power* of the system accustoms it to bear the unwholesome agent with comparative impunity." Coates continued,

> The patient is suddenly interrupted in a long continued course of hard drinking. What is then the consequence? . . . the system immediately feels the want of its accustomed narcotic. It has been gradually changed, until the

> depressing agent has become necessary to the preservation of an approach towards health; without it . . . his cerebral and nervous systems are thrown into a state of the very highest excitement.[108]

Shaping a new pathological anatomy of the inebriate, the growing medical literature concluded that habitual drunkenness had unique and disastrous physiological effects.

Disease and Panic

Tracing the influence of new theories of the mind and modes of medical investigation in shaping conceptions of pathological drinking does not fully explain the tremendous interest in alcoholic insanity, nor does it explain the direction this new inquiry took. Delirium tremens resonated within a particular socioeconomic context. As much as intellectual developments, medical interest in delirium tremens reflected a new set of professional imperatives and social concerns. With hallucinations a subject of popular fascination, delirium tremens offered a compelling subject for new forms of medical investigation. But the medical literature of the 1820s demonstrates that physicians attributed a very specific social significance to the newly prevalent disease. In the medical literature, they often told stories about delirium tremens victims that expressed anxieties consonant with the social aspirations of male medical students.

In the 1820s, educating these students had become a lucrative business. Philadelphia had two publicly chartered medical schools, including the University of Pennsylvania, and five private institutions offering medical instruction. The city also offered two hospitals and three dispensaries where students could gain clinical experience.[109] At the University of Pennsylvania alone, four to five hundred medical students attended lectures each year, making it by far the largest medical school in the country.[110] Young white men in search of medical training traveled to the city from every region of the country.[111]

Because of the emphasis on dissection in medical training, the Philadelphia Almshouse was a natural place for medical education and the practice of anatomy to flourish. The overwhelming majority of paupers entering the almshouse required medical attention and lacked the means to pay for a private physician.[112] Hoping that student fees could help fund the expenses of the medical ward, the almshouse managers built a two-hundred-seat anatomy theater in the early 1810s and also established a medical library.[113] Students paid for the privilege of gaining clinical experience and attending anatomical performances.[114] When patients in the medical ward died, most also lacked resources for burial and so became subjects for dissection. Crucially for the practice of morbid anatomy—the anatomy of disease—their medical histories were known. By the 1830s, in an imposing new building, the Philadelphia Almshouse became the largest municipal hospital in the country and the most prestigious place for medical students to gain clinical experience.[115]

Arriving in Philadelphia, aspiring physicians had great freedom in choosing their course of study.[116] In the 1820s, graduation requirements at the University of Pennsylvania included attendance at two yearlong courses of lectures (each year being exactly the same course), a three-year apprenticeship with a respectable private physician, and a written dissertation.[117] Practicing physicians in the city did take on private students, but with the influx of so many eager students, the demand quickly outstripped the number willing to devote time to examining an apprentice.[118] That said, the lack of effective licensing requirements meant doctors were not required to receive their medical degree before entering private practice.[119] As one historian has put it, "The plan of medical education that prevailed was remarkably flexible and largely voluntary, a system in which a whole host of options for professional improvement . . . were selectively pursued by physicians in accordance with their own ambitions, talents, and means."[120] Looking for ways to supplement their education, and with traditional apprenticeships in short supply, students willingly paid private instructors.[121]

Lectures remained at the heart of medical training, but students particularly valued the practical lessons offered by anatomical dissection.[122] Illustrating the morbid nature of this new mode of education, John Godman described his method of teaching as "decomposing" the body. In class sessions, students confronted untouched cadavers rather than preserved anatomical models. As the student worked, "the body is *decomposed* by the knife in his sight, and he soon acquires a clearness of information on the connexion of parts existing in the living system." Godman emphasized the importance of the experience of dissecting. "As the students *sees* the veritable anatomy for himself, his subsequent reading is always aided by recollecting the *actual condition* of the structure."[123]

In the 1820s, medical education might be characterized as democratic, in comparison with the republican medicine of Rush. Whereas Rush had carefully cultivated and defended the mystique of his professor's chair, and thus the natural hierarchical relationship between lecturer and student, in Godman's mode of education teacher and student were equals:

> By this method the *teacher* is always placed in the condition of a *learner*, and no authority is accredited by *demonstrations*—no book is valued until its descriptions have been tested by a rigid scrutiny, in direct comparison with the structure as fairly exposed, and competently observed. This appeal from *books* and *authorities* to *nature*, disperses the clouds which have too long involved the science of Anatomy—removes the difficulties that have impeded the advances of the inquiring student, and opens the way to improvement, discovery, and truth.[124]

The elevation of practical experience over the hierarchical authority of professorships, titles, and academic credentials catered to the upwardly mobile young men who crowded into the city in search of professional status.

Anatomy transformed medical education and practice, but it also introduced new modes of professionalism. Historian Michael Sappol has argued that in the antebellum period anatomy was central to "the making of American professional medical identity."[125] Physicians represented anatomical science as a reform aimed at vanquishing useless medical theorizing as well as "superstitious" popular attitudes toward the body. But because anatomy violated deep-seated social norms surrounding death and burial, the cultural power of the science was inseparable from long-standing social mores. Anatomical credentials became a central source of a new professional authority. At the heart of this claim to authority was the power derived from having entered and examined the dead body: "decomposing" it, visually confirming its inner workings, and witnessing the nature of disease and death. Citing among other things the skeleton in the doctor's office that became iconic in the late eighteenth and early nineteenth centuries, Sappol argues that doctors used anatomy to construct a death cult infused with magical authority: "Anatomy's 'charm' lay in the production of a distinctive professional charisma, a command over the body, that deliberately transgressed the boundaries between life and death, purity and contamination, and the sacred and the profane."[126]

The cultural power of anatomy helps explain why delirium tremens literature took on its peculiar gothic character. Philadelphia maintained its status as the unquestioned center of medical education in large part because the city's schools, institutes, and instructors provided students ready access to cadavers.[127] The hundreds of young men who flocked there every year in the 1820s and 1830s had opportunities to conduct dissections regularly, and at least the most diligent ones returned to the medical school every night to pursue their anatomical preparations.[128] The activities of

Philadelphia's medical students unnerved city residents. In the 1820s the medical schools' extensive use of cadavers led to a justified anxiety that grave robbers must be meeting the growing demand. By 1828 the state of Pennsylvania passed a law specifically outlawing the violation of graves.[129] Medical schools continued to obtain bodies by employing professional grave robbers under a secret agreement with the city government of Philadelphia.[130] Nevertheless, the supply of cadavers was tenuous, and conflicts periodically erupted among the city's professoriate over how bodies would be distributed. By the mid-1820s, private anatomy instructors found it harder to secure bodies in the face of the growing power of large institutions, a conflict that occasionally exploded into public view.[131]

Anxiety about the illicit appropriation of bodies elicited a range of responses.[132] The city's artisans and tradesmen formed mutual benefit organizations to build cemeteries and provide decent burials for workingmen and their families. The Machpelah Cemetery Society advertised its new burial ground with the central selling point that it offered "a Superintendant House for the protection of graves."[133] Another measure to prevent disinterment was a sturdy and secure coffin, but these were expensive.[134] In 1829 a Philadelphia newspaper advertised ready-made coffins (themselves a very recent commodity) for working people. It said in part that because of "frequently unpleasant feelings as regards disinterment," the manufacturer "has invented a preventative, which he thinks will have the desired effect, and at a trifling expense, which can at any time be examined by those who wish to purchase."[135] Occasionally anger about disinterment broke out in violence. In 1836 a rioting mob forced a church sexton to flee the city after evidence emerged that he had sold bodies from the churchyard to medical students.[136]

Medical students were not immune to this cultural antipathy toward the science of anatomy.[137] Young men entering the medical profession experienced dread and trepidation on first confronting the dead in the dissection room.[138] But many came to revel in the taboo science. One journalist in 1830 described the new generation of medical students as "death-daring aspirants" who

> wallow in the filth in the dissecting room, with a cheerful and animated countenance, and sustain the most offensive effluvia without a qualm, for the sake of unraveling the morbid condition of some rotten viscus. They will hazard their own lives to detect the cause of death in others. Nor can infection nor contagion deter them from living examination, or *post mortem* in-

> vestigation. The risk of *exhumation* [body snatching] is to them trifling, when compared to the advantages of a laboured investigation of the human frame by the dissecting knife. Their thirst for the acquisition of knowledge is as ardent and craving as the appetite of a drunkard. It is to such *spirits* as these that our profession owes its elevated rank.[139]

Sappol documents that the practice of anatomy gave rise to a fraternalist culture of "alcoholic jollity, morbid humor, dissecting-room antics, and body snatching."[140]

Interest in delirium tremens hallucinations thrived in the context of this remarkable expansion in the practice of dissection. Illustrating this connection, Isaac Snowden wrote the first American essay on the disease at the Philadelphia Almshouse just a few years after the completion of the hospital's new anatomy theater. That the apparitions of delirium tremens appeared in medical discourse at exactly this historical moment, and within this largely young and exclusively male milieu, seems understandable, especially given the growing medical literature on hallucinations and the supernatural. The horror inherent in cutting up dead bodies in the middle of the night may have heightened the appeal of alcoholic ghost stories.

The freewheeling nature of medical education during this period also created an environment where these stories thrived. The enormous demand for instruction created considerable economic opportunity as physicians competed for paying students. During this era, students paid both instructors and university professors directly for admission to lectures and demonstrations rather than paying tuition to a school. Students usually paid about $20 for a course of lectures. They also paid for the privilege of attending rounds in the almshouse and hospitals, and numerous private instructors offered a wide range of demonstrations.[141] Within the medical community, Klapp, Horner, Coates, and Gerhard were only a few of the outstanding physicians who acquired significant wealth by regularly conducting clinical lectures and demonstrations. Elite professors could attract as many as four hundred students. The annual incomes of ordinary doctors averaged $500 to $1,000, with $3,000 considered very high. For a one-semester class, a professor at the University of Pennsylvania might make $8,000 or more, apart from his private practice or private students.[142]

When doctors noted the amusing nature of delirium tremens, they implicitly made use of its theatrical potential. Graphic stories of delirium tremens hallucinations were common in lecture halls and anatomy theaters.[143] For medical instructors eager to attract paying students, delirium tremens could be exploited for economic gain. Certainly this seems true of Klapp, whose writing on the disease significantly promoted his medical career. More than any other single physician in the 1810s, he popularized the delirium tremens disease model among the Philadelphia medical community. In 1867 a prominent Philadelphia physician recalled his studies under Klapp almost fifty years earlier. He noted Klapp's popularity as a teacher and his "rare talent for imparting oral instruction, so as to render whatever he discoursed of pleasing and intelligible to his hearers, and to impress firmly upon their minds such particulars as threw light upon the pathology diagnosis, or treatment of disease."[144] Another student reminisced that Klapp took a "peculiar delight in treating" delirium tremens cases, "following them up, and describing them to his class."[145] Characterized by their lengthy descriptions of patients' hallucinations, Klapp's published essays on delirium tremens no doubt supplemented his lectures on the disorder. Significantly, in 1820, exactly the moment when his writing on delirium tremens was its most influential, and at the height of the economic hardships following the Panic of 1819, Klapp purchased a property in the southern part of the city for $20,000.[146]

Working at night in the city's anatomy rooms, doctors descended into the bloody cavities of inebriate bodies and told stories of dark, supernatural horror. In a sense, delirium tremens was a gothic performance that transformed dissection rooms, anatomy theaters, and dark basement almshouse cells into theatrical spaces. The disease allowed physicians to tell sensational tales about the phantasmagorical horrors wrought by the intemperate alcoholic imagination. Physicians used pathological anatomy to demystify alcoholic hallucinations, while their graphic descriptions highlighted the dark allure of the supernatural.

Yet for the young men who flocked to Philadelphia for medical training, as well as the instructors who took their money, delirium tremens was more than entertainment. The disease also carried a pressing significance derived from their social aspirations. Throughout the early republic, medicine represented an attractive career because of its association with gentility.[147] Whether students came of their own volition or were sent by socially ambitious parents, a medical career required a relatively low capital investment because educational requirements were minimal and flexible. Even if they had to supplement their incomes with other work, their university training, the association of the profession with European centers of learning, and membership in a national network of learned gentlemen doctors gave even cash-poor doctors a foundation for claims to a lofty social status. Whether from the northern middle class, the southern gentry, or the rural West, young men hoped that a year or two in Philadelphia would transform them into gentlemen.[148]

In this community of ambitious young men, narratives of delirium tremens turned the disease into a cautionary tale. The singular cause of this disease—habitual heavy drinking—invited commentary about personal habits. Of drunkenness, one student wrote in 1817, "Here we see man degraded in the extreme, and precipitated from that noble and dignified character and station, in which the beneficent-creator of the universe had placed him in the scale of animated beings."[149] Repeatedly in the medical literature, physicians prefaced their essays by lamenting that intemperance persisted as a common social problem. "It is indeed a cause of regret," one medical student wrote, "to see so many, who appear every way endowed by nature to fill exalted and honorable stations in society, prostitute the most splendid talents and extensive acquirements at a shrine so detestable and debasing."[150]

Commenting on the disease, these students were preoccupied with the relation between individual moral behavior and social and economic success. Doctors and students described delirium tremens as a physical and psychological disease that epitomized the horrors of intemperate drinking, failed social aspirations, and economic catastrophe. One doctor told of a patient who had been "a remarkably stout young man, and till a few weeks previous to the attack of delirium, had enjoyed uninterrupted health. He had unconsciously associated himself in business with a knave, and soon found that the small property he had accumulated was rapidly wasting." The business disappointment led him to a drinking binge, culminating in an attack of delirium tremens in which he became relentlessly haunted by "the images of a diseased imagination."[151] Physicians speculated that bankruptcy or business failure might even predispose people to the disease.[152] Doctors believed it was especially dangerous to individuals of some cultivation and learning. In 1821, one dissertation urged patience and compassion for delirium tremens patients:

> It is a fact, that a large proportion of these cases are to be met with among persons of some cultivation—sometimes among the sons of genius—still retaining a remnant of sensibility. Let us therefore bear in mind how much the heart of such an individual must have been wrung, in descending from a station of respectability and comfort, to become the degraded inhabitant of a *poor-house*. His punishment is ample, and it is our duty, in all our intercourse, to consider him as an object of sympathy, and to endeavour to soften the anguish of wounds, which we may not be permitted to heal.[153]

In describing individuals of splendid talents, cultivated minds, and glittering prospects, physicians could easily have been describing themselves or the hundreds of medical students who descended on Philadelphia each year.

The concerns expressed in this literature were quite different from Rush's republican fears. He warned that strong drink would corrupt voters, undermine their sense of social responsibility and virtue, and foster a "hatred of just government." This new generation of medical men, however, saw in delirium tremens a morality tale about masculine success and failure in the economic marketplace. Essays on the disease offered physicians the opportunity to opine on the proper habits crucial to social respectability, while their narratives of patients' sufferings dramatized the horrible consequences of falling short. One case history published in 1822 in a Philadelphia medical journal illustrated the symbolic nature of the disease in the years immediately following the Panic of 1819:

> His eyes looked wild—he kept standing, and could not be prevailed upon to be seated. He soon pointed across the room, looking very earnestly, saying to his son, "Don't you see them? Don't you see them? There they are again!" I asked him what it was he saw? He answered, *mice*, which had come to eat his library; he said they had already greased and spoiled his most valuable books and that he had sustained one hundred dollars damage by them. No circumstance of the kind had taken place. His wife informed me that he had been thus deranged four days, during which time he had not slept at all. . . . [D]uring the night, to use her own words, "he carried on in such a manner as to frighten them, and they had to call in the neighbours for assistance to manage him." I knew this man had been addicted to a free use of spirituous drink for two years; and his wife now informed me, that . . . apparently from a sense of the pernicious effect this course had upon his constitution, he suddenly abstained from its use. . . . This affection followed this sudden abstinence from his accustomed stimulus.[154]

Horrid vermin crawled out of the dark recesses of Mr. R.'s alcoholic and ungovernable imagination to despoil the most valuable objects of his library, which next to the parlor, stood as the ultimate middle-class symbol of intellectual integrity and social respectability.[155] Throughout the antebellum period, delirium tremens would carry this essential association with social downfall.

The disease thus became one important way physicians began to explain disparities of wealth in biological and psychological terms. Delirium tremens implied that failure derived from a depraved physiology. It likened bankruptcy and social downfall to a desperate struggle with "the phantoms of the ideal world, specters with gorgon heads, and bodies more hideous than those of the satyr or the fabled tenants of the lower regions." As physicians began to investigate the pathological anatomy of failure, they suggested that wealth and poverty were natural categories rooted in the science of human life.

Doctors portrayed the inebriate as not simply a sentimentalized economic failure, however, but a figure who, sunk into depravity, experienced otherworldly visions. This phantasmagorical disease helped popularize the idea that intemperance itself could be a type of theater. Even as they applied new modes of scientific investigation to the age-old problem of drinking, their descriptions of demons and other horrors were essentially gothic, betraying an undeniable fascination with the phantoms, devils, and specters their patients described. Especially in Philadelphia's medical theaters, delirium tremens offered antebellum physicians and medical students an imaginative escape. The supernatural welled up from the nether regions of the drunkard's psyche—regions lying in the deep shadows cast by the light of medical science, anatomical investigation, and human reason. As young medical students anxiously strove toward a middle-class status that was by no means assured, in the realm of the diseased imagination they found an irresistible topic of speculation.

The sweeping intervention of the medical profession into the social response to alcohol abuse thus derived from paradoxical impulses. The emergence of the delirium tremens diagnosis was at the forefront of dramatic reforms in American medicine involving a new emphasis on empiricism and scientific rigor. The disease was of central concern to the same physicians who sought to make American medical education more practical and relevant for aspiring doctors while they discarded old and discredited theories. Delirium tremens demonstrates that American physicians were closely following developments in the European centers of medical learning and were eager to bring new theories to the American profession. Pathological anatomy sought to demystify the effects of liquor on human physiology and to find visual evidence of the seat of alcoholic insanity. But the interest in delirium tremens also mystified and romanticized the inebriate. As pathological drinking became a distinct category of medical knowledge, physicians demonstrated their interest in a disease not of the will, as Rush had called it, but of the imagination. The new diagnosis became an amusing spectacle and an opportunity to moralize about the common causes and frightening consequences of failure. Elite instructors and their students took voyeuristic pleasure in contemplating habitual drunkenness as the antithesis of the imperatives that drove their own tenuous social aspirations, betraying a lurid fascination with the victims of hard drinking, downfall, and despair.

THREE

Hard Drinking and Want

When he embarked for London in 1800, James Oldden Jr. was just nineteen years old. The oldest son of a wealthy Philadelphia merchant, he traveled to meet family and friends and to cultivate the connections crucial to his future success in business. The extensive diary he kept reveals a young man eagerly aspiring to genteel society. Summing up his experiences in England, Oldden wrote, "The comforts, luxuries, and conveniences of life I have enjoyed to their fullest extent—been kindly received, & entertained by number of those friends to whom I have made myself known."[1] When his youthful travels ended, he set the diary aside for nearly a quarter of a century.

In 1824 Oldden felt compelled to pick it up again and write one last entry summing up the business career that had followed his European tour. It reads as a melancholy postscript to what had been a narrative of confident ambition. Beginning in his twenties, he had struggled with painful rheumatism, seeking relief to no avail. His commercial pursuits had also been difficult, something he blamed on "the changes of times." As he borrowed large sums and invested in risky shipments to foreign markets, his occasional successes were followed by larger failures. In 1817 Oldden turned to the coffee trade, hoping to emerge from his increasingly desperate financial circumstances. A small success and modest fortune led him to a disastrous investment: "If any bounds could have been set by my transported imagination, I should now have been saved the painful recital of a scene of suffering & woe. . . . I might have come out with a handsome profit. But *no* again I must outstrip all reason, all prudence & buy nearly 1,000,000 pounds—a sudden fall took place."[2] The drop in coffee prices ruined him.

Oldden gave himself up to drink. For the next six months he "indulged copiously, and oh the additional agony it produced. Vilest of all vile evils that afflict the world what cause have I to look upon it with horror." Alcohol abuse led him to the medical and emotional crisis that prompted him to remorsefully pick up his diary again:

> My mind . . . gradually gave way to melancholy, which was greatly increased by the recourse I made to stimulus & going on from one step to another I became at last so lost to all sense of propriety, as to use from a quart to three pints of Ardent Spirits every day—this produced temporary insanity, & my friends by the advice of a Physician sent me twice to the Pennsylvania Hospital.[3]

Hospital records confirm that Oldden was admitted on April 27 and July 10, 1823, for "mania a potu." Despite retaining friendships with such prominent men as John Quincy Adams, Oldden never regained the social standing he enjoyed as a young man.[4] He also continued to struggle with drinking. Throughout the 1820s, the formerly wealthy merchant listed himself in the city directory under the more modest occupation of "dealer." The hospital again admitted him for mania a potu in 1827 and 1832, now poignantly noting his occupation in admission records as "late merchant," casting his social downfall as a sort of death.[5] Oldden died in 1832, less than two weeks after being released from the hospital. The administration of his estate recorded that he had just purchased a new biography of the enormously wealthy Philadelphia merchant Stephen Girard.[6] Respected Philadelphians celebrated Girard's life as a shining example of the broadly held belief that business success and social advancement were open to all men who worked hard and aggressively pursued opportunity.[7] Oldden's career stands as a stark contrast to Girard's success.

Failed merchants who turned to drink had been well-known figures in the eighteenth century. What distinguished Oldden's nineteenth-century narrative was his diagnosis of mania a potu. In his diary, Oldden did not use this name to describe his affliction, likely because in 1823 few outside medical circles knew the term. Delirium tremens had only very recently become a common diagnosis at the Pennsylvania Hospital, and discussion of the disease remained largely confined to medical wards, lecture halls, professional journals, and student dissertations. In that discussion, however, doctors saw an important social significance to this new disease. Beginning in the tumultuous years surrounding the financial Panic of 1819, physicians commonly asserted that men of sensibility, education, and cultivation were more prone to delirium tremens. Oldden's narrative about the role drinking played in his descent into penury mirrored the stories that doctors and students told each other. In medical journals and dissertations, doctors reported that these patients were most often "individuals of cultivated minds, lofty sentiments, and glittering prospects," and "the sons of genius."[8] Especially, they were like Oldden: men who had turned to drink because of business disappointments, bankruptcy, and failure.

That intemperance and social downfall commonly go hand in hand was a powerful truism in the early American republic, part of the understanding that a man's wealth and social status were mainly of his own making. When writing the story of his troubled business career, Oldden dwelled on his personal failings, giving weight and attention to his excessive drinking, but he largely neglected the numerous historical developments that had created a treacherous business climate, such as the Embargo of 1807, the War of 1812, and the Panic of 1819. He was far from the only merchant, temperate or not, to fail during these troubled years.[9] It is also noteworthy that despite suffering from painful rheumatism, Oldden primarily associated his destructive drinking with his failures in business. Alcohol would have been the cheapest and most accessible option for pain relief in the early nineteenth century. When Oldden turned to his diary to narrate his life, however, he chose to blame his drinking for the loss of his cherished station as a Philadelphia merchant. For both physicians and patients, the cultural linkage of intemperate habits and social downfall was compelling.

Was Oldden typical of delirium tremens patients? Did the concerns physicians expressed in their writing on the disease accurately reflect its epidemiology? Searching for answers to these questions illustrates that new medical responses to alcohol abuse developed in a social context of capitalist transformation and economic instability. Physicians' intense interest in delirium tremens occurred as part of a conversation about intemperance, poverty, and respectability. Developing class distinctions, along with other categories of social difference such as gender and race, shaped both perceptions of heavy drinking and the experiences of those who suffered its consequences.

The stories doctors told about delirium tremens influenced how they applied the diagnosis. After 1819, medical and municipal records began to count people suffering from the disease as a distinct category of patient. Inebriates thus became visible and quantifiable as never before. When we cobble together what survives of these records, a portrait emerges of how physicians and city officials interpreted the troubling social and health consequences of alcohol abuse. Oldden's experience of alcoholic insanity exemplifies these changes. He was diagnosed with a disease, put under the care of a physician, treated in a hospital bed, and recorded in the patient register. City institutions did not give this relatively privileged treatment to common drunkards. If they were given any attention or shelter at all, they were confined to almshouse cells. Physicians thus sought to respond medically to behavior linked with the social downfall of respectable men.

By midcentury, alcohol abuse remained pervasive in an increasingly fractured society. Into the 1830s and 1840s, commentators, reformers, and leading citizens blamed intemperance for a host of social ills such as urban poverty, criminality, depravity, and evil. In Philadelphia, however, this angry conviction never accurately reflected the social makeup of problem drinkers. As intemperance carried an ever heavier stigma, respectable patients who suffered from delirium tremens continued to seek treatment in the city's medical institutions. Physicians treated the habitual drinking of worthy patients as a medical problem while ignoring the poor and unfortunate. Through the delirium tremens diagnosis, they sought to distinguish the troubling fall of middle-class inebriates like James Oldden from the depravity of the faceless and intemperate poor. As the decades wore on, the close association physicians had seen between delirium tremens and the "sons of genius" faded, and the disease became more of a depressing daily reality than a cutting-edge diagnosis. But the contradictions that had inspired fascination with the delirium tremens diagnosis only became more pressing as drinking and failure continued to haunt the socially ambitious middle class. In American medicine, society, and culture, this diagnosis thus signaled new responses to alcohol abuse that both reflected and shaped widening disparities of wealth and class in the early republic.

Interpreting Poverty and Failure

In the 1810s, as Philadelphia physicians were first treating and studying delirium tremens, city leaders debated how to respond to rapidly growing poverty. Long-simmering concerns about the cost of poor relief boiled over into outrage about the perceived drinking habits of the poor. A series of events combined to create the crisis on city streets. Dislocations during the War of 1812 brought refugees into the city. A particularly harsh winter in 1816–17, combined with a spike in fuel and food prices, caused widespread suffering. Following soon after, the Panic of 1819 bankrupted citizens of all social classes and threw thousands out of work. A destructive riot in the fall of 1819 at the city's Vauxhall Gardens and an outbreak of yellow fever during the summer of 1820 further contributed to a sense of gloom.[10] Throughout these years, the ongoing shift by the city's manufacturers toward a system of wage labor made workers far more vulnerable to seasonal fluctuations in employment or other economic disruptions.[11] In the years after 1815, the number of persons entering the almshouse increased by 80 percent.[12]

City residents had long worried about the problem of poverty. When prominent citizens confronted the crisis of the 1810s, their ideas were shaped by the legacies of controversies dating back to the 1760s and the economic crisis that followed the Seven Years' War. A central point of these debates had been to what degree the poor were personally responsible for their condition. Philadelphia's most famous resident, Benjamin Franklin, had suggested that the best way to alleviate poverty was to eliminate poor relief. The dependent would thus have more incentive to improve themselves.[13] Many prosperous citizens also felt that the poor were unwilling to work and should be compelled to be more industrious. The city had long maintained a small almshouse, but it became overcrowded in the economic recession of the 1760s. Influential citizens, with Franklin again a key leader, raised funds for a larger, reorganized almshouse that came to be called the "Bettering House."[14] When it opened in 1767, the institution's administrators compelled the able-bodied poor to labor at menial jobs such as weaving or cobbling shoes. Promoters championed this system as a way to instill new habits of industry in the poor, giving them the discipline they needed to become useful citizens. Put to profitable use, inmates' labor might also alleviate the financial burden poor relief placed on Philadelphia taxpayers.

Philadelphians did distinguish between the idle poor and the industrious or "worthy" poor, who had come to their condition through no fault of their own. Because injury and disease could have a devastating financial impact, the Pennsylvania Hospital also served as a key institution in Philadelphia's system of poor relief. In the 1750s, Franklin led the effort to found the hospital, which served to preserve the worthy poor from moral contamination by the degenerate population at the almshouse.[15] Benjamin Rush was key to further expanding medical care for the indigent in 1786 when he joined other reformers to found the Philadelphia Dispensary. The dispensary provided only outpatient care, with the rationale that patients would benefit from recuperating at home rather than in an impersonal hospital. In 1802, the dispensary treated some 17,500 patients.[16] The system of poor relief that survived into the nineteenth century included the almshouse, the Pennsylvania Hospital, the Philadelphia Dispensary, and a system of "outdoor relief" that provided food, fuel, and small cash payments to the needy outside the almshouse.

The transformation of the city's almshouse into a "Bettering House" failed to stop the swelling of the ranks of the poor because it did not take into account the underlying reasons for poverty, which derived from economic dislocation.[17] The belief that poverty was a personal failing continued to shape the city's poor relief services.[18] During the winter crisis in 1817, leading Philadelphians formed the Pennsylvania Society for the Promotion of Public Economy to help those who were suffering. As part of their mission, the committee set out to "investigate the causes which contribute to produce the deplorable number of individuals, who annually require public charitable assistance."[19] A committee circulated a list of eighteen questions to administrators of relief agencies and other influential citizens, including several leading physicians. The committee's questions were simple, including "What description of persons are most improvident?" and "How many children can an industrious husband and wife support by daily labour?" The committee then published its findings, including its own summary with direct quotations from respondents.

The findings were striking—many testified to the structural nature of poverty. Did husbands commonly abandon their families and leave them impoverished? The committee wrote, "Few persons, we presume, are prepared to anticipate the result of this inquiry; the evil is lamentably extensive." The committee refuted the "popular opinion" that blacks benefited disproportionately from public assistance as "not supported by the facts derived from the documents." Perhaps most significant, the committee concluded that wage laborers and those in seasonal trades, such as carpenters and plasterers, often lacked the resources to support themselves during the winter. On the straightforward question of whether the poor would work if they had the opportunity, however, the committee reported great difference of opinion. "They will all say they are willing to work, but the fact is not so," reported one man. But others disagreed: "Many of the poor are willing to labour if they could procure employment."[20]

These sharp disagreements illustrate the failure of leading citizens to confront the human consequences of economic change. The committee did agree on one conclusion. "We have no doubt," a typical respondent wrote, "but the immoderate use of ardent spirits is the principal cause of poverty . . . and many of those who receive public charity expend a part in procuring that article." In the midst of the harsh winter, another wrote simply, "The great cause of suffering is the intemperate use of ardent spirits."[21] This idea was not new, but the years following the report saw the link between intemperance and poverty become a point of broad consensus among leading citizens. In the context of an unstable economy and growing poverty, the heavy use of liquor became a principal explanation for troubling social transformations.

The economic depression that followed the Panic of 1819 exacerbated the crisis surrounding poverty, but it also visited suffering on a broader cross section of society.[22] While economic flux was not new to Philadelphia, the depth and scope of the downturn were unprecedented. Since the turn of the century, the expanding availability of credit and the growing reliance on paper currency had created new webs of commercial interdependence. These new market structures had been the source of great optimism during the postwar economic boom, but they were also highly unstable.[23] In 1819 the fragile banking and credit system collapsed, creating a wave of spectacular bankruptcies and foreclosures. Unlike previous crises, the depression affected every sector of the economy simultaneously. As one newspaper editor reported, "agriculture, industry, and commerce alike feel the pressure."[24]

At the height of the crisis, Philadelphians struggled to quantify the suffering. A citizens' committee survey of manufacturing between 1816 and 1819 found that employment had dropped almost 80 percent, a figure later confirmed by the 1820 federal Census. Tens of thousands of people, both workers and their dependents, lacked any means of support.[25] Over 1,800 individuals languished in Philadelphia jails for unpaid debts, ranging from as little as $10 to upward of $40,000.[26] And those incarcerated represented only a fraction of insolvents, since not all creditors chose to jail their delinquent debtors. The number of persons seeking some form of long-term relief far surpassed any previous period of economic upheaval in the city. In 1819, 4,049 individuals entered the almshouse, a number that would not be surpassed until 1846, when the population of the city had grown by 250 percent.[27]

The scale of the crisis intensified the debate regarding the nature of poverty and the best way to provide relief services even after the depression subsided in 1823.[28] In 1828, hardening attitudes toward the poor led Philadelphia to become the only large American city to abolish all forms of public outdoor relief. Now the city required that all recipients of poor relief be confined to the almshouse. The same year, the city completed an enormous new almshouse in Blockley Township, on the west side of the Schuylkill River, far from Philadelphia's city center.

Apart from the debate over poor relief, much published writing during the Panic of 1819 struggles with a troubling question: How could one explain the failure of respectable, hardworking men? The crisis marked the painful inauguration of the boom-and-bust cycle that would characterize the market economy taking shape in the nineteenth century.[29] The catastrophe struck Philadelphia's middle class of merchants, entrepreneurs, bankers, and manufacturers as hard as any group, since their economic fortunes had been tied into the fragile banking system.[30] In 1819 the city's newspapers commonly reprinted anecdotes describing remarkable bankruptcies, such as a man "who failed for more than half a million, whose private wine vault, as it stood at the time of his bankruptcy, was estimated to have cost him 7000 dollars! We heard of another who failed lately—the furniture of one suit of apartments on a single floor, cost $40,000."[31] Commentators were shocked that failure seemed to touch men regardless of their character, habits, or station. A newspaper correspondent professed a disbelief that was common: "Men whose honor is proverbial, and whose credit stood as high as any in the world, are prostrated and bankrupt, and the affliction pervades every avenue of society."[32] Another newspaper correspondent asked, "Do we not see ruin, and misery and want, preface every rank of private life?"[33]

It was in the midst of this ruinous economic crisis that physicians began to associate delirium tremens with failures and bankrupts like James Olden. While city leaders had arrived at a consensus that intemperance was to blame for growing poverty, in published case histories and student dissertations physicians were far more sympathetic to delirium tremens victims. They commonly connected alcohol abuse with economic concerns. Writing in 1823, one medical student lamented the social consequences of "the immense consumption of spirituous liquors." He wrote, "Our almshouses, and hospitals are filled with its votaries, and even in our streets we daily behold its numerous victims dragging out a miserable existence in penury, and want."[34] The victims of delirium tremens were not simply poor, however. The disease was "most commonly met with among those who have once enjoyed the comforts and luxuries of life."[35] Case histories often offered explanations for why these patients had fallen on hard times. "Their business and the duties of office have plunged one man into frequent hard drinking while cares and misfortunes have goaded on another. . . . Here genius and talent are leveled with the dust, in trying to forget in wine, the outrages of fortune, and the ingratitude of the world."[36] Physicians' sympathetic narratives echoed sentiments expressed in newspapers about honorable men now "prostrate." They also contrasted sharply with the condemnation of the intemperate depravity of the poor.

The delirium tremens diagnosis marked patients as distinct from common drunkards. Doctors recommended that these patients be separated from the general patient population, urging that "kind and affectionate treatment be extended towards this unfortunate class of people, not only because it is most consistent with the best principles of humanity, but also as the most successful means of restraining their passions and effecting their recovery."[37] Writing about delirium tremens patients, one medical student stressed that

> the feelings of the spectator will be often severely tried when he sees patients who have once enjoyed all the ease, luxuries which wealth can bestow treated as though they possessed not a spark of human nature. It will generally be found that though from the long, and habitual use of strong drinks the sensibility of the patient may appear to be lost; yet he always possesses enough to appreciate the conduct of his keepers. In most of our public institutions too little attention is paid to the proper selection of persons to take charge of the patients; and they are thus left too frequently to the brutal ignorance of men totally unfit for the charge.[38]

A delirium tremens diagnosis meant these patients were more deserving of benevolent treatment.

In the years immediately following 1819, delirium tremens thus came to powerfully evoke the hard consequences of failure in an unpredictable economy.[39] The association of drinking habits and failure also had eighteenth-century roots. Benjamin Franklin's popular writing, most famously his *Autobiography*, epitomized the American self-made man, offering a set of ideals that modeled masculine behavior for the socially ambitious. Franklin championed the importance of cultivating a respectable public persona as a path to economic success and social advancement. Young men could build character by cultivating frugality, industry, and temperance through rigorous adherence to self-improvement. In part an expression of the republican culture of the early national period, the cult of the self-made man was popularized in a burgeoning advice literature for young men.[40] The corollary to this American myth of opportunity was that failure resulted from weakness of character, and that intemperance was the most common cause of all. All young men had limitless potential, this literature taught, because character could be formed by hard work, education, and moral values. Social success was the inevitable result of temperance and industry. But the malleability of the young left them open to pernicious influences.[41] The social landscape created by the volatile economy heightened these fears. For the young men who embraced Franklin's American promise, the Panic of 1819 demonstrated that a sudden fall into poverty could happen to anyone.

These ideas were intensely felt among Philadelphia's young doctors and medical students, in part because of the distinctive nature of their professional training. The ambitious flocked to Philadelphia's rapidly expanding medical schools because the profession represented a chance for bourgeois respectability. Their training, however, required them to work in institutions, such as the Pennsylvania Hospital, the Philadelphia Almshouse, and the city's dispensaries, that primarily served the poorest and most unfortunate members of society. In striving for success, these young men and their professors confronted misery, misfortune, and failure every day. In delirium tremens they saw a frightening disease that epitomized the dangers threatening their own social aspirations.

Drinking and Death

Immediately after the Panic of 1819, these cultural associations that linked drinking habits with the economic fate of individuals began reshaping responses to alcohol abuse along lines of social class. Even as delirium tremens became a common diagnosis, city institutions also preserved the older and more general label "intemperance." Physicians, administrators, and coroners used "intemperance" to refer to intoxication or more generally to a patient's habits, and they cited it as a major factor in a plethora of health problems. The patient records at the Pennsylvania Hospital reveal, for instance, that the hospital rarely admitted patients suffering from intemperance but commonly treated delirium tremens, a disease caused solely by habitual heavy drinking. Individuals overcome by intoxication were most often taken to the almshouse, where both "intemperance" and "delirium tremens" appeared regularly in inmate records.

Mortality statistics and burial records kept by the Philadelphia Board of Health show when this new distinction took shape.[42] The Board of Health had long tracked "intemperance" and "insanity" as causes of death, but in 1821 it included the new category "mania a potu."[43] Reported deaths from mania a potu rose rapidly during the decade and peaked in 1832, when 150 deaths were attributed to the disease citywide. These statistics must be understood in the context of an urban population that grew from approximately 70,000 in 1800 to over 388,000 by 1850.[44] Figure 4 traces deaths by "intemperance," "insanity/mania," and "mania a potu" as an overall percentage of mortality in Philadelphia between 1808 and 1840.[45] At its peak, a disease that lurked invisibly before 1821 accounted for close to 3 percent of all mortality in 1833.

Figure 4 raises a number of difficult questions. Why the spike in deaths

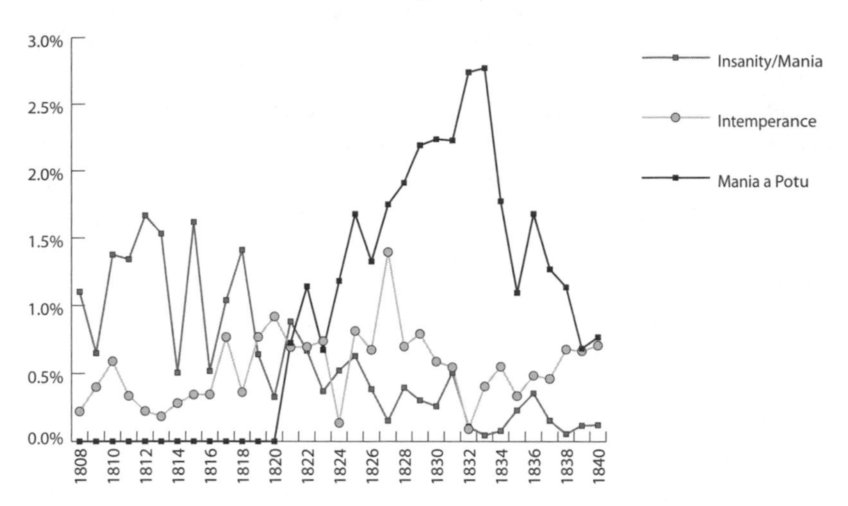

Figure 4. Deaths by intemperance, insanity, and mania a potu. Calculated from the Philadelphia Board of Health's published Statements of Deaths, collected in Susan Klepp, *"The Swift Progress of Population": A Documentary and Bibliographic Study of Philadelphia's Growth, 1642–1859* (Philadelphia: American Philosophical Society, 1991).

in 1832 and 1833, followed by a precipitous decline? As early as 1824, the *Philadelphia Gazette* called this increase in deaths from mania a potu "astonishing."[46] The spike demonstrates that physicians and city officials came to "see" alcohol abuse as a cause of death far more often than ever before. Not only did mania a potu far outstrip intemperance and insanity as a cause of death, but the inclusion of the disease in mortality statistics marked a vast expansion of deaths attributed to alcohol abuse and a distinct decline in those attributed to insanity. A committee of physicians writing in 1831 attempted to calculate the overall number of deaths in the city attributable to alcohol abuse. Taking into account all the adverse health consequences of drinking, they estimated that more than 15 percent of annual mortality resulted from heavy drinking.[47] And just as striking as the increase in deaths attributed to mania a potu is their precipitous fall. By 1840, mania a potu and intemperance combined accounted for the same percentage of overall mortality as they did in the early 1820s.

Do the mortality statistics correspond to incidence of delirium tremens in the city? Given the haphazard and fragmentary nature of the medical records that survive from this period, it is difficult to extrapolate more generally about the prevalence of the disease. Surviving records offer sharply contradictory information. The Philadelphia Almshouse physician W. W. Gerhard claimed, for instance, that between 1834 and 1839, the number of cases annually admitted to the medical ward more than doubled, from 121 to 274, while deaths from the disease plummeted from 75 in 1832 to just 3 in 1841. Gerhard trumpeted this statistic in defense of his particular treatment strategy.[48] The Philadelphia County Medical Society later dismissed his statistics as "extravagant," implying that he had intentionally inflated the number of patients admitted with delirium tremens to puff up his reputation.[49] On the other hand, Pennsylvania Hospital treated virtually the same number of delirium tremens cases in the 1830s as it did in the 1840s, and the mortality rate remained relatively constant.[50] Overall, the evidence suggests the incidence of delirium tremens was relatively constant. The disease remained common in Philadelphia medical institutions throughout these years, and by midcentury, mortality rates improved.

Why did the Board of Health list deaths from intemperance and mania a potu in separate categories? Given that the sole cause of delirium tremens was habitual heavy drinking, a death from delirium tremens was by definition a death from intemperance. Surviving medical and municipal records demonstrate that this categorization reflected physicians' understanding of delirium tremens as a disease of social downfall. In addition to compiling the aggregate mortality statistics that figure 4 is drawn from, the Board of Health also documented "cemetery returns," or individual internments.[51] A survey of these records yielded the names, ages, cemeteries, doctors, and dates of death for 907 individuals who died of delirium tremens and 634 who died of intemperance between 1825 and 1850. Taken together, these two groups represent a majority of individuals who, according to the Board of Health's more inclusive mortality statistics, had died of some form of alcohol abuse.[52]

Information on these 1,541 people is scarce, but the evidence that survives demonstrates clear differences between these two groups along the lines of gender, race, and class. Someone described as having died of intemperance was much more likely to be poor; 62 percent of intemperance victims were buried in either the almshouse or "city public" cemetery rather than in a church or private burial ground. Less than a third of delirium tremens victims were buried in paupers' graves. In the 1840s, the cemetery returns began including phrases such as "intemperance and want" and "intemperance and exposure" as causes of death, making the link with poverty more explicit. Gender also distinguished the two categories. Overall, women made up less than one-third of the group of 1,541 individuals.[53] However, delirium tremens victims were overwhelmingly male, while women made up nearly half of the people who succumbed to intemperance.[54] Race appears inconsistently in the cemetery returns, but the records show that delirium tremens victims were overwhelmingly white. Between 1844 and 1850, 67 African Americans died of intemperance, compared with only 12 who died of delirium tremens. Race, gender, and class were much more important factors than age. These individuals tended to be in the prime of adult life; men were, on average, thirty-eight years old when they died of delirium tremens, and women were two years younger. On average, individuals dying of intemperance were several years older. The average age for women who died of intemperance was thirty-eight, and for men, forty-two.

The social stigma associated with habitual drinking contributed to making delirium tremens victims overwhelmingly male. The *Philadelphia Gazette* asserted that the recorded deaths from delirium tremens "are believed to be but a part of those which occur. From delicacy, physicians generally give the disease a milder name, and thus make it difficult to ascertain the whole extent of the malady."[55] One instance in the records of the Northern Dispensary illustrates this possibility. A Mr. and Mrs. Story were both treated at the dispensary on January 3, 1845.[56] The titles Mr. and Mrs. rarely appear in the records, suggesting that this couple enjoyed a higher social standing than common dispensary patients. While the register of patients lists Mr. Story as suffering from delirium tremens, Mrs. Story's complaint appears as "nervous fever," a common euphemism used for delirium tremens when trying to protect the patient's reputation.[57] Mrs. Story died of her affliction, and her husband survived. What Mr. and Mrs. Story suffered from and how often medical records employed euphemisms for delirium tremens is impossible to know.

Temperance supporters believed it was common. An often reprinted letter from a physician to a temperance society claimed that "respectable families are often afflicted by members who fall a sacrifice to their indulgence in spirituous liquors; and the physician who prepares a [death] certificate of the case cannot employ the disgraceful and shocking terms '*Drunkenness, or Mania a Potu*,' which would . . . deeply wound the feelings of a family already sufficiently distressed. He is therefore obliged to call the disease, '*Inflammation of the brain, Insanity, &c.*'"[58] That poor women make up a higher percentage of intemperance deaths suggests that propriety played less of a role in classifying the deaths of impoverished individuals.[59]

Depraved Paupers

Do these municipal records confirm the censure of prominent citizens and commentators? Were the poor intemperate? By historical standards the answer is likely yes, but this was a period of very heavy drinking among Americans in general. Historian W. J. Rorabaugh has estimated that alcohol consumption rose precipitously in the early nineteenth century. By 1830 the average American over age fifteen drank an estimated seven gallons of alcohol annually, approximately three times the current rate. This was made possible by widespread availability and affordability: in the early nineteenth century, whiskey cost less than "beer, wine, coffee, tea, or milk."[60]

Despite the temperance education campaigns of Benjamin Rush, many people in the 1820s still understood alcohol to be generally healthful. Especially during harsh winters, cheap distilled liquor likely provided some modicum of comfort for people unable to afford proper food or sufficient fuel to keep warm, and whiskey was affordable for all.[61] What is impossible to know is how the drinking habits of the poor compared with those of the prosperous citizens who condemned them. Without private homes to retreat to, the poor were (as they are today) far more likely to drink in public and so were much more exposed to public censure than wealthier classes. A poor person overcome by intoxication in public risked incarceration at the almshouse, where the incident would have been recorded. Heavy drinkers with private homes were less likely to face this risk. Even if they needed medical attention, drinkers of even modest means could call a doctor to treat them at home and so were less likely to rely on services from a public institution or to leave any historical record of the incident.[62]

The most significant factor differentiating people recorded as dying of intemperance versus delirium tremens was the circumstances of death. The coroner certified at least 60 percent of the intemperance deaths recorded in the cemetery returns, while physicians certified almost all deaths from delirium tremens. The coroner investigated and certified deaths from ambiguous causes, most commonly those of poor people who died outside a doctor's care. When a death occurred, the coroner convened a jury of common citizens to inquire into the circumstances. One Philadelphia author described the arbitrary and melancholy nature of the inquests to dramatize the suffering of the poor at the hands of indifferent city officials. The jury, he wrote, could have been "selected as representatives of the phases of degradation which give to cities distinctive classes—embodiments of the active and the passive vices. The 'loafer,' the brawler, the sharper, and the 'pot-house' politician were there; each, in every word and gesture, showing forth the distinctive characteristic of his class." Investigating the tragic death of a baby, the "deliberations of the jury were conducted with all the haste and indifference with which men dispatch unimportant affairs—for what is so unimportant as the reason why death should overtake a pauper's child."[63]

Published coroner's inquests from the 1830s and 1840s in the Board of Health records demonstrate that death from intemperance said more about the social position of the deceased than about their drinking habits. Intemperance was one of the coroner's most common verdicts. In 1831 the longtime Philadelphia coroner John Dennis wrote in a letter to a temperance society that his experience had left him with a firm conviction that the use of ardent spirits caused premature death: "I have no hesitation whatever in avowing it as my firm belief, that the use of intoxicating liquors is the prolific cause of a great proportion of the deaths which come under my view as Coroner."[64] Brief stories of some of the individuals the coroner certified as dying of intemperance can be found in the *Public Ledger*.[65] This daily newspaper began publishing select coroner's inquests after 1836. Reading just a few of the published inquests confirms that the coroner's use of intemperance as a cause of death was little more than a moralistic condemnation that assigned personal responsibility for privation, suffering, and death. One account described a woman "who was found dead in her bed with a bottle containing liquor beside her. It appears that her husband left about half past six o'clock in the morning, and it is supposed from the state of the body when found that she had been dead several hours. . . . At the time of the inquest, her husband had not been found . . . Verdict, death from intemperance."[66]

Poverty typifies these narratives, and in many cases "intemperance" may explain death from exposure or starvation. Published in the midst of the economic depression following the Panic of 1837, the inquest into the death of a forty-year-old black woman reported that she was found dead in the morning in a shed in a lot below Carpenter Street, between Third and Fourth Streets. "She was . . . an habitual drunkard. Verdict, death from intemperance. Her husband was with her at the time of her death; they both having been ejected by their landlord from their dwelling, took refuge in the hovel where she was found dead."[67] In ruling on the cause of death, the coroner privileges her reported drinking habits. Other possible factors caused by her abject poverty, such as exposure, malnutrition, or disease, go unmentioned.

The high percentage of women who died of intemperance relative to delirium tremens partly reflects the demographic makeup of the poor. In antebellum Philadelphia, women were less likely to be admitted to the almshouse or Pennsylvania Hospital, but they nevertheless made up a majority of the poor.[68] The coroner was thus more likely than a physician to certify the deaths of poor women, because they tended to die outside an institution or a doctor's care, in uncertain circumstances. In the cemetery returns sample, 50 percent of the individuals the coroner judged to have died of intemperance were women.[69] The few spare case descriptions that survive in newspapers starkly dramatize how the term was used as a moral condemnation of marginal women in morally suspect circumstances. "Intemperance," for instance, might refer to spousal abuse. Mary Ann Ward appears in the cemetery returns as having died of intemperance during the winter of 1849. The published coroner's inquest reads, "The Coroner . . . held an inquest on the body of Mary Ann Welsh, alias Mary Ann Ward, who died the evening previous, in Broad street, between South and Shippen. Death from intemperance and exposure, in connection with abuse at the hands of Thomas Welsh."[70] "Intemperance" could also sum up the death of a prostitute. A newspaper item commenting on the death of a twenty-one-year-old woman read, "Death from Intemperance—A wretched prostitute, whose real name is Jane Colder, but who has usually gone by the name of Elizabeth Kinsman, died on Wednesday morning, from the fruitful cause of all vice and misery, intemperance. To this cause may be attributed her fall and course of life—from this cause she met her death."[71]

The coroner's use of "intemperance" in assigning a cause of death thus tells us little if anything about the habits of these unfortunate people. Instead, the term served to inscribe social difference. For the Philadelphia coroner, personal habits and social class were the same thing. In the case of Levi Lee, published in the winter of 1847, the coroner's judgment of death by intemperance seemed to have no connection to the circumstances:

> On Thursday night, while the snow storm was at its height, Levi Lee, a colored man was found dead on the steps leading from a cellar in the rear of Spafford street below Shippen. The deceased had been an occupant of the cellar . . . having been sick for some time, the suspicion arose that he had the fever which has prevailed . . . among this wretched class of persons. This induced the woman to order the sick man out, which on his refusing, she is reported to have beat him most brutally and finally pushed him out into the steps to where he was found dead.

At least in the published account, there is no mention of alcohol. But for the coroner, "the conclusion was irresistible that the unfortunate man died from intemperance and exposure, and so the jury decided by their verdict."[72] "Intemperance" stigmatized Lee, assigning him personal responsibility for his suffering and death and resolving the persistence of African Americans, extreme poverty, and infectious disease in the city. The wretched were intemperate by definition.

At no time were these attitudes about intemperance and poverty more in evidence than during the 1832 cholera epidemic. During the first half of the year, the city's newspapers printed terrifying accounts of the "Indian cholera" as it spread through Asia into Europe. "Its spread is awful," read one newspaper item, "for in the short space of fourteen years it has desolated the fairest portions of the globe, and swept off at least *fifty millions* of human beings."[73] As it spread from one European city to the next, it often appeared first in poor neighborhoods. In printed accounts, commentators associated the disease with the filth of the lowest classes. Cholera was a poor man's plague. By summer the disease had finally jumped across the Atlantic to Canada and spread throughout the United States along the recently constructed networks of roads and canals.[74]

In American cities, prevalent attitudes toward poverty and class shaped perceptions of the disease. A permanent underclass characterized European cities, but as Charles Rosenberg describes, Americans denied that their cities possessed the same ossified social disparities. In Philadelphia, cholera thus became a disease closely associated with depraved personal habits, and especially with drunkards. The frightening prospect of its spread further inflamed anger about the perceived habits of the poor. Already cause for disgust, drunkards now threatened to infect the city with a deadly disease. One physician wrote to a Philadelphia newspaper that "the disease has searched out the haunts of the drunkard and has seldom left it without bearing away its victim." An 1832 Philadelphia broadside stated the simple principle,

QUIT DRAM DRINKING
If you would not have the
CHOLERA.

This transfiguration of the disease from a poor man's plague to a disease of depraved habits was reinforced by the Philadelphia physicians who rallied to combat the affliction. Under the section heading "The classes of people, and the modes of living of those who have died in the greatest numbers," a report on cholera commissioned by the College of Physicians of Philadelphia strongly recommended that "abstinence from ardent spirits at all times desirable, is, in seasons of pestilential visitation still more necessary." Although unable to decide whether cholera was contagious, the commission had little doubt that intemperance contributed to its spread. Surviving records from an 1832 cholera hospital show that patients' drinking habits were carefully noted. This linkage of alcohol abuse to cholera was not inconsistent with contemporary medical theory. Nevertheless, the single-minded emphasis on ardent spirits spoke to deep concerns about the continued growth of poverty in the city.[75]

Why did the spike in delirium tremens deaths, illustrated in figure 4, coincide with the cholera epidemic? Unfortunately, the Board of Health's records offer no help, since they contain only thirteen of the mania a potu deaths that occurred in 1832 and 1833.[76] The weekly totals the Board of Health published in 1832 show that the deaths from delirium tremens occurred throughout the year, rather than bunching during the month of the epidemic.[77] However, the simultaneous drop in the number of intemperance deaths in 1832 suggests one explanation for the spike. Cholera dominated the city's attention throughout the year. Especially because of Philadelphia's constant maritime traffic, physicians and public officials watched nervously for any signs of the disease. A broad term, the coroner's diagnosis of intemperance would not have ruled out cholera. The spike in mania a potu deaths, then, likely resulted from increased medical surveillance of the poor, as medical and municipal officials scrutinized the causes of mortality.

Cholera cemented the connection between wretched poverty and the depravity of intemperance in Philadelphia's public discourse. Through the 1830s and 1840s, newspaper items related to intemperance became even more dark and disturbing. Newspaper editors linked heavy drinking not just to poverty and failure, but also to suicide, poverty, gambling, theft, riot, murder, infidelity, prostitution, corruption, and all other forms of illegality, depravity, violence, and tragedy. These items' regular appearance sounded a steady drumbeat repeatedly linking liquor with social and moral evil. Familiar headlines screamed "DREADFUL EFFECTS OF INTEMPERANCE," "INTEMPERANCE AND DEATH," and "HORRIBLE EFFECTS OF INTEMPERANCE—MURDER OF A YOUTH BY HIS GRANDFATHER."[78] When one woman committed suicide while incarcerated in the watch house, the only explanation the *Aurora* offered for her death was that she had been intoxicated during the evening:

SHOCKING END OF INTEMPERANCE.

> We learn that at an early hour of the night before last, a woman named Mary White, was taken up in the street, much intoxicated and put into the back room of the Watch House. On yesterday morning, at the time of releasing the prisoners, one of the officers remarked that a woman who had been put in, had not come out, and on making examination found a lifeless body. It appeared that she had tied her handkerchief to one of the bars in the window, and fastened her garter so as to form a loop, with which she had hung

herself. The deceased left 7 children, who unlike their wretched mother, are all well behaved, and placed in good employment.[79]

The "good behavior" of the innocent children contrasted with White's suicidal intemperance.

Newspaper accounts portrayed intemperance as a monstrous evil. Even the most hideous atrocities were attributed to drunkenness. Commonly the family home was the scene of terrible alcoholic dramas. Only masculine effort, self-restraint, and industry, along with self-sacrificing female devotion to domestic responsibilities, these accounts insisted, could keep the demon rum from invading the fragile hearth and home. In an item titled "Intemperance!" the *Saturday Courier* reported. "It would be utterly impossible for any journalist to keep an account of the awful murders that are perpetrated from the use of intoxicating drink!" George Collins, the article continued, shot and killed his oldest son and severely wounded another of his sons while in a fit of temporary insanity brought on by habitual drinking. Offering "the use of intoxicating drink" as the primary explanation for horrid crimes, narratives made liquor into a powerful metaphor for evil.

An Affliction That "Pervades All Ranks of Life"

While newspapers dwelled on these sensational accounts of poverty, suffering, and radical evil, physicians and medical students working in city institutions knew that intemperance was a far more complicated social problem. Physicians' early medical literature had cast the "sons of genius" as the most susceptible to delirium tremens, and in medical journals some physicians did continue to associate the disease with middle-class failure, but the social reality of delirium tremens was more diverse. In an extended essay on delirium tremens published in a Philadelphia weekly newspaper in 1835, a Philadelphia physician asserted that intemperance "is not confined to any particular class; it pervades all ranks of life—the rich and the poor—the learned and the ignorant the rude and the polished; and among the victims of intemperance may be found the man of genius, talent and learning as well as the stupid and the less endowed."[80] Medical records from 1825 to 1850 confirm that, while overwhelmingly white and male, people judged to have died of delirium tremens came from a broad cross section of society. On average, however, those dying of delirium tremens were more affluent and included a higher percentage of middling and wealthy individuals than those judged to have died of intemperance.[81]

While it is one thing to say that victims of delirium tremens came from "all ranks of life," it is another to explore what that statement really means. After 1800, conceptions of social rank and class were undergoing a fundamental transformation. One of the most important new social fault lines was the divergence of manual and nonmanual occupations. Factories, textile mills, and other industries rapidly turned Philadelphia from a port city focused on Atlantic world commerce into one of the leading centers of manufacturing. This transformed the work lives and economic prospects of ordinary Philadelphians. The changes affected different groups at different times. With the spread of wage labor, journeymen, mechanics, apprentices and other manual workers faced declining economic fortunes, so that by 1850 the average male manual worker was making only three-fifths of the income necessary to support an average-sized family.[82] At the same time, a new middle class emerged as the market economy spawned a growing number of nonmanual jobs in a broad range of occupations, including retail merchants, clerks, bookkeepers, and salesmen.[83] Others worked in elite professions such as law and medicine, while master artisans reorganized their workshops around capitalist forms of production or moved into retailing. The income of these nonmanual middle-class workers grew in the middle third of the century, even as the unstable economy exposed them to considerable uncertainty. Occupations and economic prospects formed the basis of new class identities as the daily experience of manual and nonmanual workers diverged in family organization, consumption, and outlook.[84]

How temperate were members of the new middle class? Did they adhere to the maxims popularized in Benjamin Franklin's mass-printed *Autobiography*? By 1850, as drinking became increasingly unacceptable socially, Rorabaugh estimates that Americans on average came to drink dramatically less, probably less than half the 1830 rate. Widely cited by historians, Rorabaugh's estimates address national consumption, however, and as such they do not reveal how drinking patterns may have differed in various regions of the country. In the 1840s, for instance, social commentators often lamented that while America's small towns had become mostly temperate, heavy drinking remained very common both in large cities and in western frontier areas. It is unclear how alcohol consumption in a large and diverse city like Philadelphia might compare with a city like Rochester, New York, which was deeply influenced by evangelical temperance reformers. What is more evident is that fermented and distilled beverages increasingly took on strong symbolism. While cheap whiskey became increasingly associated with poverty, depravity, and disease, wealthy elites began cultivating wine connoisseurship and outlining strict rules for wine-drinking comportment to convey gentility, wealth, and refinement.[85]

In midcentury Philadelphia, medical records show that heavy drinking nevertheless remained a ubiquitous, intractable, and troublesome social problem that showed no signs of abating even among the respectable.[86] Most individuals who died of alcohol abuse came from manual occupations, but many also came from middle-class professions.[87] Exactly how many is hard to determine. During this period, many men listed their occupations under traditional manual categories but may have transitioned to becoming retailers, dealers, or other nonmanual middle-class professions. An individual identified in the directories as "hatter" or "boots" for instance, may have been a wage-earning hat or boot maker or, increasingly as the century wore on, a middle-class retailer of hats or boots. Despite these difficulties, doctors working in antebellum hospitals confirmed that many in the middle class were drinking heavily. Pliny Earle, a physician at the Bloomingdale Asylum in New York City, compiled the occupations of delirium tremens patients admitted between 1821 and 1844. He highlighted that "merchants, traders, clerks, professional men, persons of leisure, and young men without employment" made up more than half the patients treated for delirium tremens. "The necessary inference," he said, "must be far from flattering to these classes."[88] Earle qualified the observation by noting that the patient population of the Bloomingdale Asylum was more affluent than those of other institutions.

As Earle suggested for New York, in Philadelphia social class, not just wealth, was a factor in determining where inebriates received treatment. Of the four hundred individuals treated for delirium tremens at the Pennsylvania Hospital between 1829 and 1849, approximately one-third listed nonmanual occupations in their admission records, including thirty merchants, seventeen clerks, ten attorneys, four doctors, and four manufacturers. Because of its founding mission to treat the worthy poor, the hospital accepted cases seen as immoral in nature, such as delirium tremens, only if the patient paid a weekly fee. For delirium tremens, the charge ranged from $3 to $5, although fees sometimes went much higher. Admitted for a rare case of "intoxication," one "gentleman" paid $12 a week. While overall the hospital admitted more than half of its patients free, it received payment for 546 out of 557 cases of delirium tremens.

The occupational diversity of delirium tremens victims is striking. Of the 1,541 people in Philadelphia's cemetery returns, the occupations of 219 can be found in the city directories. From carders, clerks, and constables to coopers, grocers, and carpenters, these individuals worked in ninety different types of employment. Occupation can give some idea of material wealth.[89] More than half of these 219 people were employed in middling artisan occupations such as butcher, carpenter, hatter, and printer. Another 29 came from occupations that offered opportunities for a comfortable living, such as silversmith, piano maker, and coach maker, and 17 listed occupations identified with the wealthy, including gentleman, attorney, and merchant.[90] By comparison, only 41 individuals who died of intemperance could be found in the city directories at all, and almost half of those came from occupational categories that yielded bare subsistence or less, such as mariners, washers, and weavers.

Of these 219 people, if any single occupation stands out as especially prone to heavy drinking, it is tavern keeping. The physician Daniel Drake's wrote of tavern keepers:

> Many intemperate men resort to these occupations as furnishing the means of support and indulgence, when all other resources are dried up; but . . . it is undeniable, that an extraordinary proportion of these people become drunkards, from the custom of drinking hourly, with those who frequent their establishments. Thus he who is an accessory to the suicide of so many others, becomes eventually the destroyer of his own life.[91]

Philadelphia's records show that Drake's characterization likely was true. Tavern keepers and innkeepers make up 10 percent of those found in the city directories. Some owned well-established businesses. Andrew Bossart's Seven Stars Tavern had been in operation at the corner of Fourth and Race Street for twenty-five years when he died of delirium tremens in 1839. The tavern contained eleven well-appointed rooms, with the beds alone valued at $414.[92] James Maher's tavern was more modest, with five rooms for patrons crowded with cots and mattresses. The value of Maher's entire estate was just $179.[93]

Popular culture commonly warned of the horrors of drinking and promised that hard work and self-restraint offered a sure path to economic and social well-being. Working at the Philadelphia Almshouse Hospital, however, physicians and medical students commonly saw patients who had been arbitrarily victimized by disease, injury, and economic misfortune. These people had complicated personal histories that could not easily be reduced to the harsh condemnations of drunkards that circulated, for instance, in temperance literature, coroner's inquests, or newspaper items. Most of the people who died of delirium tremens while at the almshouse either had been admitted for reasons unrelated to alcohol abuse or faced difficult circumstances that had little relation to their drinking habits.

Much of the evidence pertaining to inebriates at the almshouse is anecdotal. The fragmentary records that survive from the hospital (called the Philadelphia Hospital by the late 1830s) cannot provide statistical measures comparable to those from the Pennsylvania Hospital.[94] The records that survive, for instance, do not always include a diagnosis. A resident physician lamented in 1844, "Who would ever imagine that a death-book afforded the only records of a large hospital, yet such is the fact; and it is only of late that any diagnosis of disease has been registered on the patient's admission."[95] Using other municipal records, however, it is possible to identify a large group of people who died at the almshouse of alcohol abuse. The survey of the Board of Health cemetery returns found 442 individuals who died of either intemperance or delirium tremens and were buried in the almshouse's burial ground.[96] The dates provided by the survey of the cemetery returns make it possible to locate a large number of inebriates in surviving almshouse records.[97]

Although both hospitals catered primarily to the poor, between 1825 and 1850 delirium tremens patients at the Almshouse Hospital were markedly poorer than those at the Pennsylvania Hospital. Of men with occupations noted, 92 percent were manual workers, unskilled laborers, and paupers. Half this group had been born in the United States, and a fifth of those were native-born Philadelphians. Considering the demographic makeup of the city, immigrants made up a disproportionately large percentage. For instance, almost one-third of the group were Irish immigrants. In 1830, by comparison, foreign-born immigrants made up just 10 percent of Philadelphia's overall population.[98] The Pennsylvania Hospital also had a large percentage of immigrant patients. Just under a quarter of the inebriates treated there were Irish. In both hospitals, English and German immigrants were tiny minorities. The percentage of Irish immigrants in both institutions remained constant throughout the 1830s and 1840s. Given these numbers, it is interesting that physicians only rarely associated delirium tremens with Irish immigrants.[99] The high percentage of Irish patients in these two institutions indicates more about their socioeconomic status, which compelled them to seek help in city institutions.

Almshouse records show that most people who died of either delirium tremens or intemperance had recently been the victims of misfortune. For instance, the loss of a spouse preceded 25 percent of deaths from alcohol abuse. Because employers excluded women from all but the lowest-paid employment and male manual workers earned too little to support a family, a spouse's death could be financially devastating. By contrast, only 7 percent of the patients at the Pennsylvania Hospital reported being widowed.[100] Even for middle-class families of some means, the death of a spouse could begin a quick slide into dependence. Mary Craig died of "intemperance" in the almshouse at age seventy-two. An Irish immigrant, Mary came to the United States in her twenties with her husband, who eventually bought a six-hundred-acre farm in rural Pennsylvania. After her husband died, she moved to Philadelphia and entered the almshouse in the winter of 1827. She died there in May 1828.[101] Elizabeth Niket, the widow of a Philadelphia jeweler, was twenty-three when she entered the almshouse and died of "intemperate use of opium and ardent spirits." Her husband had died earlier the same year.[102]

Few details survive about inebriates' broken families and orphaned children. At both the almshouse and Pennsylvania Hospital, approximately 45 percent of inebriates were married at the time of their admission. At the almshouse, most who were either married or widowed left from one to eight children. A forty-four-year-old rope maker, Jacob Miller, lost his wife a year before he died of delirium tremens. He left five children, but no mention is made of their ages or where they went after their father's death.[103]

Complicated medical histories often preceded an early death by delirium tremens. Illness and injury were the most common reasons individuals entered the almshouse: 85 percent of new inmates required medical attention.[104] At least 40 percent of inebriates in the cemetery returns sample visited the almshouse more than once, most often for medical attention, and as many as 50 percent of those who died there of delirium tremens had been admitted for unrelated complaints such as "dropsy," "frostbitten," and "dog bite." James Furlow, described as a twenty-six-year-old "equestrian" of intemperate habits, was admitted after being injured in a fall. He died several days later of delirium tremens.[105] Isaac Wheater was an English weaver, a widower who had worked in Philadelphia for many years. In 1838 his leg was amputated at the Pennsylvania Hospital. A year later, he was confined at the almshouse for two months and put to work winding bobbins. In the summer of 1841, he died of delirium tremens.[106]

Common almshouse stories like Isaac Wheater's suggest that many patients resorted to heavy drinking to deal with painful medical conditions. Indeed, for many this coping strategy may have begun during hospital stays, although we cannot know this for certain, since we do not know their drinking habits before entering the hospital. Physicians commonly prescribed alcohol, however, and often in large quantities. Published accounts for 1817 demonstrate that the Almshouse Hospital purchased over 1,200 gallons of whiskey, in addition to large quantities of brandy, porter, and wine, for the 1,800 patients it admitted that year.[107] Physicians did commonly use laudanum, a tincture of opium, but liquor remained a medical mainstay throughout this period. At the almshouse in the late 1830s, liquor became the primary treatment for delirium tremens after years of experimenting with emetics and opium.[108] Liquor was also without question the cheapest and most readily available painkiller to anyone who was outside an institution and could not afford a doctor's care.

Some physicians worried that giving patients large amounts of alcohol created a poisonous habit. Writing in 1829, the Pennsylvania Hospital physician and temperance activist Benjamin H. Coates asked: "Are not physicians in various ways the means of introducing habits of intoxication? Are they not, in too numerous instances, instrumental in leading their patients into this destructive practice, by the long continued use of these substances as a medicine?"[109]

The Daily Occurrence Docket offers evidence that Coates's fears were well founded. Hester Parker, for instance, an unmarried twenty-five-year-old black woman, spent the first six months of 1841 in the almshouse venereal ward. Readmitted in September of the same year, she spent another six months recovering from an injured leg. Three weeks after being discharged, she was readmitted with pneumonia and died of delirium tremens.[110] Edward Maxwell, a thirty-one-year-old factory worker, had been in Philadelphia just one week when he broke his arm in an accident on the street. Released after three months, Maxwell was readmitted just two months later and died of delirium tremens, leaving a wife and four children.[111]

The medical histories of almshouse patients thus demonstrate that the circumstances of their deaths challenged a middle-class ideology that saw poverty as a moral failing and attributed success to industrious habits. Most often these patients had been victimized by events outside their control. Appearing in newspapers, temperance literature, and other forms of popular print, the simplistic narratives that treated impoverished drunkards as little more than filth did not correspond to the reality of alcohol abuse in the city.

Failed Inebriates

This tension between ideology and social reality grew by midcentury. As Philadelphia's population neared 400,000, disparities of wealth and class became more and more pronounced.[112] Drinking and temperance became intensely symbolic of social distinctions, but the victims of alcohol abuse continued to come from across the social spectrum. John Daley, a carter, died in 1839 of delirium tremens with an estate valued at $42. His horse and cart accounted for $22 of that wealth. His possessions consisted of a copper kettle, a looking glass, seven chairs, a stove, a "lot of prints," clothes, one bed, three tables, a washstand, a desk, and miscellaneous items such as tins and crockery. Daley was married and a father. The 1830 census counted him as the head of a fifteen-person household. Likely made up of multiple families, the household included six children under age five, two teenagers, and seven adults.[113] Far more prosperous men also suffered the consequences of intemperate habits. William Keim was a thirty-five-year-old plumber when he died of delirium tremens in 1844. With an estate valued at $4,800, his administration shows he lived in a large two-story home with a carpeted entryway, stairway, and parlor filled with consumer goods, including a case of stuffed birds, marble lamps, and a $30 parlor carpet. The 1840 census suggests that Keim's family, consisting of him and his wife and three girls under age ten, was typical of the mid-nineteenth-century middle-class household.[114]

If alcohol abuse persisted among the respectable classes at midcentury, so too did failure. The Panic of 1837 again dramatically illustrated the fragile nature of America's market economy as, over several years, the interdependent web of small financial institutions collapsed in waves of defaults. Bank failures, insolvency, and deflation ruined thousands of American entrepreneurs, with the economy coming almost to a standstill. Even the liquor business suffered. In 1842 one businessman reported that cash was so scarce in Indianapolis and St. Louis that liquor could not be sold. Demand was so low in New York City that a barrel of whiskey was selling for just 16½ cents.[115] The crushing depression illustrated for many that failure was an endemic part of the market economy. By the Civil War, it was conventional wisdom that ninety-five out of a hundred business ventures ended in bankruptcy.[116]

At city institutions, physicians and medical students confronted the arbitrary nature of failure, disease, and death every day and thus witnessed the tensions between ideological explanations of success and the social reality it purported to explain. Writing on delirium tremens, physicians saw the problems of alcohol abuse and failure as the same. "How truly a melancholy spectacle it is to behold those who might have been ornaments to their country and useful members of society giving up themselves to the intoxicating draught," wrote one young medical student in his dissertation on delirium tremens.[117] Had alcohol abuse been the domain solely of the wretched and depraved, physicians would never have seen inebriates as an important subject of study and treatment. In responding to delirium tremens, they professed concern primarily with worthwhile individuals who had brought on their own destruction.

Physicians treated even very prosperous members of middle-class society for alcohol abuse. Perhaps the most striking illustration was the death in 1836 from delerium tremens of Francis R. Godey, brother of Louis A. Godey, publisher of *Godey's Lady's Book*. Issued in various forms from 1830 to 1898, the magazine was edited for forty years by Sarah J. Hale and stood as the most popular women's magazine of the nineteenth century.[118] Filled with sentimental literature, moral instruction, and fashion tips, it epitomized the literary and consumer culture that grew up around middle-class ideals of domesticity and "true womanhood." Francis Godey's funeral was held at Louis's home.[119] Other prominent Philadelphia men included Edward Whitely, publisher of the annual Philadelphia city directory. He died in 1823 just weeks after being treated for delirium tremens at the Pennsylvania Hospital.[120] A private physician repeatedly treated William Spragg, publisher of Philadelphia's weekly newspaper the *Saturday Courier*, for delirium tremens. He died insolvent in 1843.[121]

Some delirium tremens victims had personal connections with the physicians who published articles on the disease. In 1836, apothecary Charles Nancrede died of delirium tremens at the almshouse, despite belonging to a wealthy French family. At the time of his death, two of his brothers were distinguished physicians in the city. Joseph Nancrede had written a medical journal article on delirium tremens in 1818.[122] Thomas Sully Jr. died in 1847 of delirium tremens at the Pennsylvania Hospital. His famous father had painted portraits of Dr. Joseph Klapp and his wife in 1814.[123] More than any other individual, Klapp had popularized the delirium tremens diagnosis among Philadelphia doctors.

Taken together, the diversity of the 1,541 individuals found in Philadelphia's cemetery returns thus illustrates the social basis for changing medical responses to alcohol abuse. Indeed, these individuals demonstrate why the inebriate became a subject of medical study. Physicians accepted the unregenerate depravity of drunkards and almshouse denizens: men like Patrick Cain. The sailor was repeatedly confined to the almshouse for drunkenness and destitution. When he died in 1829 at age forty-one, the administrators of his estimated $20 estate were three of his creditors, including a tavern keeper, a distiller, and a grocer, likely attempting to collect liquor debts.[124] But an ascendant middle-class ethos provided no easy explanation for why a relatively wealthy young clerk and war veteran like William Hansell would give himself up to drink. Hansell was twenty-nine in 1848 when he died of delirium tremens. His estate included close to $1,000 in cash and $1,700 in real estate.[125]

In treating alcohol abuse as a medical problem, physicians thus grappled with the nature of wealth, social status, and failure in a context of rapid socioeconomic change. In inebriate wards, physicians met members of their own social class and even individuals they were well acquainted with. "What shall be done for the Victims of Intemperance in the bosom of our own families, and if not of our own, of those of our friends?" one physician asked in arguing for the asylum. "If any one will survey, carefully, the circle of his own acquaintance," wrote another, "he will scarcely fail to find one or more families whose peace is destroyed and whose fair hopes concerning some beloved son, or father, or brother, are blasted by the curse of intemperance."[126]

If alcohol abuse and failure persisted into the 1840s, why the decline in reported deaths from delirium tremens around 1840, illustrated in figure 4? Improving therapies are one possible explanation for the decline, although there is no convincing evidence. The Pennsylvania Hospital has the only continuous patient records from the period. It treated virtually the same number of delirium tremens cases in the 1830s as it did in the 1840s, and the mortality rate declined very little, from 14 percent to 11 percent.[127] It is possible that the mortality rate improved at the almshouse. In the late 1830s and 1840s, the almshouse abandoned "Klapp's cure," which called for inducing violent vomiting. Instead, physicians there treated the affliction with opium and brandy.[128] But treatment strategies employed in the city as a whole remained diverse. Physicians often debated the relative merits of opium, emetics, alcohol, animal magnetism, and, after midcentury, ether and chloroform, but they did not reach a consensus.[129]

More likely, the decline in reported deaths from delirium tremens reflected the shifting perceptions of the disease. Delirium tremens remained a common diagnosis, but in newspapers, temperance lectures, and popular novels, the disease had become a subject of lurid speculation. Sensational accounts commonly featured in newspapers, gothic fiction, theater, and the histrionic temperance lectures associated with the boisterous Washingtonian movement of the 1840s.[130] Especially through narratives that closely associated the disease with radical evil and shocking depravity, popular culture exacerbated the stigma it had always carried. By the 1840s, then, delirium tremens no longer bore the meaning it had following the Panic of 1819, when it was a progressive medical diagnosis. Delirium tremens was no longer a prestigious subject of medical investigation and instead had become a dreary routine in large hospital inebriate wards.[131] The shifting significance of the disease may have made physicians more reluctant to cite delirium tremens as the principal cause of death, even if it still occurred daily in hospital inebriate wards.

What is clear is that by midcentury a diagnosis of delirium tremens no longer mitigated the social stigma of intemperance. In the 1840s, as the new middle class came to be a distinct social group, physicians began searching for new ways to interpret and respond to a scourge that persisted among their families and friends. But physicians were just as concerned about the health consequences of alcohol abuse as they were about preserving the social status of those afflicted. Betraying these concerns, one physician wrote: "Many a father's hopes have been blasted, and many a mother's heart torn with anguish, when a cherished son has returned from college, or from an apprenticeship, or from foreign travel with the mark of the destroyer upon him."[132]

In the context of growing disparities of wealth and the emergence of social classes with divergent economic prospects, alcohol abuse challenged ideological explanations for emerging class distinctions. As delirium tremens lost its original meaning, physicians sought new ways to respond to the class tensions bound up in a death from hard drinking.

FOUR

The Benevolent Empire of Medicine

"At this hour [on July 4, 1830], throughout the wide extended range of American territory," Dr. Benjamin H. Coates began, "its thronging population is assembled to celebrate the triumph of our political independence." A wealthy and influential physician, Coates had traveled to Philadelphia's outlying industrial neighborhood of Kensington to describe the dangers of drink for a meeting of the Kensington Young Men's Temperance Society. "We have met," he continued, "not to listen to accounts of the victories of man over those who would enslave him, not even to the arguments of patriotism for the extension of knowledge or the augmentation of productive industry." Rather, "We have come here to urge upon one another and upon our [fellow] citizens the habits which tend to preserve those faculties from premature imbecility and untimely dissolution."

Coates had been a founder of the society. In his professional life, he had also devoted considerable energy to studying the health consequences of intemperance and responding to them. Only a few years earlier, he had published an influential article on delirium tremens in a leading medical journal, and he had written extensively on the medical consequences of heavy drinking. No American doctor was more qualified to declare, "Misery and humiliation are in the cup and death lurks behind the bowl." Any physician "who witnesses the banquet of the intemperate," Coates continued, immediately recognizes "the awful presages of long and painful disease, of the distress of families, and at a distance somewhat more remote, of slow but unfailing mortality."[1]

When Coates gave this speech, temperance had become the most popular social reform movement in a historical period sometimes called the Era of Reform.[2] Temperance societies had been percolating in the Northeast since the first decade of the nineteenth century. Small and local, they were led by wealthy citizens and religious conservatives concerned about moral decay brought on by social and economic upheaval.[3] But in the late 1820s temperance had become a national movement. Founded in Boston in 1826, the American Temperance Society coordinated a nationwide upwelling of organizations dedicated to alerting people, especially young men, to the dangers of drinking. By 1829, just a few years after its founding, the ATS claimed close to a thousand affiliate societies nationwide.[4] In the following decades societies continued to proliferate, growing ever more diverse in their membership and goals. Temperance activism became a constant presence in American life and, by the 1850s, a potent political force.[5]

How did the temperance movement achieve this popularity and influence? Historians have tended to focus on two important groups: Christian evangelicals and successful entrepreneurs.[6] The founders of the American Temperance Society were in part responding to a call sounded by the eminent Congregational minister Lyman Beecher. In 1827, in the service of his larger effort to broaden the influence of Christianity in American life, Beecher published *Six Sermons on . . . Intemperance*. "Intemperance is the sin of our land," he thundered, "and, with our boundless prosperity, is coming in upon us like a flood; and if any thing shall defeat the hopes of the world, which hang upon our experiment of civil liberty, it is that river of fire which is rolling through the land, destroying the vital air and extending around an atmosphere of death."[7]

Beecher's book turned out to be the most influential publication on temperance since Benjamin Rush's *Inquiry into the Use of Ardent Spirits*. Religious activists nationwide heeded Beecher's call and began supporting temperance activism. In Philadelphia, evangelical Presbyterians were the first to respond. From the pulpit of the First Presbyterian Church, evangelical minister Albert Barnes urged his congregants to embrace conversion and temperance.[8]

Very wealthy businessmen provided much of the financial backing for temperance societies. These men were most often employers who had a financial stake in a sober and industrious workforce. Additionally, many of the young men who joined temperance movements either worked in or aspired to careers in business. The appeal of temperance to this group was twofold. First, temperance offered a new mode of comportment compatible with the emerging world of white-collar work. Abstaining from drink became a badge of respectability for ambitious young men anxious to present themselves as industrious and moral, traits thought to be crucial for economic and social success. Second, employers, especially those engaged in new capitalist forms of production, needed a more disciplined workforce to maximize systems of mass production.[9] Industrialists hired evangelical lecturers to preach the blessings of temperance to workers on shop floors. In Philadelphia, for instance, Matthias Baldwin was one of the wealthiest and most powerful entrepreneurs to support the temperance movement. He was also an apostle of industrial capitalism. After beginning as an apprentice jewelry maker, in 1825 Baldwin opened a machine shop that grew into the massive Baldwin Locomotive Works. He was the quintessential self-made man, and a heroic statue of him still stands in front of Philadelphia's city hall. For Baldwin, evangelical religion and temperance shaped an ethic of self-improvement that was the basis of his own success and represented an ideal model of discipline for the men who labored in his machine shop.[10]

As organizers, writers, and speakers, physicians were among the most persuasive figures in the temperance movement. Philadelphia was a national center for temperance activism and the heart of an extensive network of physician activists. Doctors dominated the Pennsylvania Society for Discouraging the Use of Ardent Spirits. In 1831, for instance, a vice president and twelve of the thirty-seven managers of the society were physicians, and only six were ministers.[11] These physician activists saw themselves as following in the hallowed footsteps of Benjamin Rush. "Many of the ingenious theories of the teacher are passing into oblivion," wrote the Pennsylvania Society in an anniversary message, "But, as an early and strenuous advocate for temperance . . . his name can never be forgotten, nor his worth overshadowed."[12] Philadelphia physicians published nationally circulating popular health journals that championed temperate drinking habits. They were also regular contributors to the many temperance newspapers, journals, magazines, and pamphlets produced and distributed by the interlocking national network of reformers known as the "benevolent empire." Nationwide, the most notable temperance doctors had received their medical training and graduated from the University of Pennsylvania, including Daniel Drake (1816), Reuben D. Mussey (1809), George Hayward (1812), Charles D. Meigs (1817), and Thomas Sewall (1811).

Among temperance advocates, the promise of better health and the threat of early death constituted a constant theme, even of evangelicals. Ministers may have had the drunkard's eternal soul foremost in mind, but they based their appeals as much on earthly dangers as on the eternal ramifications of drunkenness.[13] Beecher's *Six Sermons*, for instance, spent considerable time detailing the "host of bodily infirmities and diseases" associated with intemperance, including "loss of appetite, nausea at the stomach, disordered bile, obstructions of the liver, jaundice, dropsy, hoarseness of voice, coughs, consumptions, rheumatic pains, epilepsy, gout, colic, palsy, apoplexy, insanity," and other afflictions. In accordance with contemporary physiological theory, Beecher focused particularly on liquor's deleterious effects on the stomach.[14]

The growing popular conviction that temperate drinking habits were healthy was a historical achievement of the medical profession. Exemplified by Coates, doctors espoused the physical and mental benefits of abstaining from drink while drawing a stark portrait of the health consequences of intemperance.[15] While firmly rooted in contemporary medical science, however, physicians' appeals for temperance were inseparable from prevailing middle-class attitudes about race, gender, wealth, and poverty. One measure of the centrality of health concerns to nineteenth-century temperance ideology was the well-known two-print series by Currier and Ives *The Tree of Intemperance* (1849) and *The Tree of Temperance* (1872) (figs. 5 and 6). Withered and dying, the Tree of Intemperance bears immoral fruits that include lying, the almshouse, idiocy, and the wrath of God, while the trunk of the tree represents disease, misery, and insanity. Under its gnarled and thinning branches lies a depraved urban underworld on one side and a despairing rural family living in squalor on the other. By contrast, the strong branches of the Tree of Temperance shelter a white middle-class family on the way to church and an orderly and fertile farmstead. Piety, morality, and industry are among the many fruits gracing the tree, and the trunk that supports the thriving branches is health and strength of body. Physician activists like Coates shaped this enduring belief that physical and mental health was fundamental to social respectability and economic prosperity.

Why did physicians take an interest in the temperance movement? Why did Coates feel compelled to rouse himself from his comfortable home at Seventh and Walnut on July 4th and travel across town to the Kensington district to address young workingmen on their drinking habits? The son of a wealthy Quaker merchant and philanthropist, by 1830 Coates had secured a coveted position at the Pennsylvania Hospital. He was not a practicing Quaker or an evangelical, and while certainly ambitious, he was hardly self-made. His considerable wealth derived from his birth. Further, Coates had no employees or any financial stake in Philadelphia's rapidly growing industry. Given his wealth, respected standing in the medical profession, and secure future, what motivated him?

Studying physicians' commitment to anti-alcohol activism contributes to a broader understanding of both temperance reform and the developing professional identity of American doctors. By the mid-nineteenth century, the temperance cause held a compelling appeal for a broad cross section

Figure 5. Nathaniel Currier, *The Tree of Intemperance* (New York: N. Currier, 1849). John Hay Library, Brown University.

of Americans. Medical claims about the benefits of temperance and the dangers of intemperance proved crucial to the movement's national popularity.[16] Appealing especially to the socially ambitious and aimed explicitly at the middle class, the temperance impulse in American medicine was grounded in scientific claims about universal principles of human health that applied equally to rich and poor and was not linked to specific forms of Christian piety. In an increasingly fractured society, the appeal of this

Figure 6. Nathaniel Currier, *The Tree of Temperance* (New York: Currier and Ives, ca. 1872). Library of Congress Prints and Photographs Division.

medical temperance literature blurred lines of social difference by grounding social inequality in human physiology.

For Coates and other prominent physicians, temperance activism was a response to the pressures of the medical marketplace.[17] No doubt, as Coates explained in his 1830 speech, physicians' support of the temperance movement grew in part from new medical responses to alcohol abuse. For almost half a century, since Benjamin Rush launched his education campaign in 1784, medical students at the University of Pennsylvania had studied the health dangers of heavy drinking. Given this enduring interest, medical practitioners' involvement in the temperance movement was as natural as their working in the cholera hospitals during the epidemic of 1832. In his lecture, Coates spoke from considerable knowledge and experience, born largely of Philadelphia doctors' intense interest in delirium tremens.[18] Temperance organizations provided ready platforms for demonstrating professional knowledge gained from treating heavy drinkers.

Temperance societies also offered American doctors a new sense of purpose and meaning. Temperance supplied aspiring physicians with a new mode of professionalism, which proved useful in winning patients' confidence and building a private practice in a competitive marketplace. Working for the cause let them cast their profession in terms of republican egalitarianism and liberal ambition. Physicians marketed themselves as more than healers of individual patients: they were central to the health and well-being of society. Claiming leadership in the national temperance movement let doctors assert professional authority over what many citizens perceived as the most pressing social issue of the day. In contrast to the earlier writings of Benjamin Rush, however, after the 1820s physicians' collective sense of social responsibility was balanced with a stronger appeal to individual self-interest. In speeches, pamphlets, essays, and popular journals, they translated advanced medical science into a helpful program of self-improvement that equated physical and mental health with economic prosperity and social advancement.

The New Medicine of Self-Improvement

In the 1820s, temperance activism was one expression of the revolution in American medicine. The physicians who championed temperance were for the most part young and ambitious. The careers of some of the most distinguished physician activists in Philadelphia, including Coates, David F. Condie, John Bell, and Charles D. Meigs, illustrate how physicians' involvement in the temperance crusade grew out of a larger campaign to reform medical practice and education. Coates, Condie, Meigs, and Bell were colleagues, roughly the same age, all born between 1792 and 1797. They attended the University of Pennsylvania together. Meigs and Bell graduated in 1817, Condie and Coates in 1818. These four young men aspired not just to be physicians but to attain the upper echelons of the fiercely competitive Philadelphia medical establishment. Toward this end, they positioned themselves at the forefront of the scientific movements that were transforming medical education and practice as well as the way the profession represented itself to the public. The most important intellectual movements in this transformation were new French theories of physiology and the science of phrenology.

Coates, Bell, and Condie became the leading promoters of the theories of physiology and phrenology within the Philadelphia medical community. Although phrenology and physiology had different intellectual histories, they shared important common assumptions. Both asserted that "all the phenomena of animated nature are displayed through organization."[19] The physiologists followed this idea in studying internal anatomy through postmortem dissection, while the phrenologists studied the mental faculties by measuring the structure of the skull. Framed as the disinterested description of human and animal nature, both movements also made sweeping claims about their egalitarian potential. According to their advocates, knowledge of phrenology and physiology would empower men and women to better themselves. Physical appearance, health, morality, and social status could all be shaped by a person's habits of body and mind. The path to self-improvement was industry, education, and above all temperance. Physiology and phrenology would form the intellectual basis for physicians' commitment to the temperance crusade, but both movements began well before the flowering of Philadelphia's temperance societies.

In 1822 Coates and Bell founded the Philadelphia Phrenological Society, the first such society in the United States. Its mission was to disseminate ideas previously confined to elite medical and intellectual circles.[20] Along with Coates and Bell, the society included some of the city's leading physicians, including Philip Syng Physick, then one of the grand old men of Philadelphia medicine, and the young William E. Horner, who would go on to dominate the teaching of anatomy in the city in the late 1820s and 1830s. The Phrenological Society also attracted prominent citizens including wealthy lawyers and merchants. The society laid the groundwork for the enduring popularity of phrenology in Philadelphia.[21] When the famous British phrenologist George Combe toured the United States in the late 1830s, attendance at his lectures in Philadelphia far outstripped attendance in other American cities.[22]

Phrenologists believed that the structure of the skull corresponded to the relative strength or weakness of the mind's various faculties or capacities. The Viennese physician Franz Joseph Gall had developed the science in 1796, and knowledge of it circulated in American medical circles soon after.[23] The Philadelphia Phrenological Society sought to use this brain science to establish the most successful modes of self-improvement. Founded in the middle of the roughly four-year economic depression that followed the Panic of 1819, the concerns of the Phrenological Society reflected widespread anxiety. The depression had created unprecedented distress and poverty and inspired an outpouring of concern about the depraved habits of the poor.[24] In the context of economic misery, Bell described phrenology as having "an important bearing on our social happiness."[25] Phrenologists hoped that deducing individuals' intellectual and moral strengths by examining their skulls would help them "direct, with a prospect of success, that great moral engine, Education."[26] "Each man has his gifts," Bell wrote, "and he should be cautious how he attempt what neither nature or education give him the power to accomplish."[27]

But the society's activities were not strictly utilitarian. Coates wrote several lectures that suggest it also catered to the literary and scientific pretensions of the city's elite. His 1824 lecture on the organ of "Ideality," for instance, sought to analyze phrenologically how writing poetry engaged all the mental faculties. In particular, Coates rhapsodized about Homer's *Odyssey,* asserting that the descent of Ulysses into hell represented the highest expression of the specific faculty of ideality. For Coates, Homer's description of "the darkness, the shadows, the silence, and the dampness of the tomb; . . . surrounded with indistinct terrors, and peopled with the shades of the departed existence" best expressed the operation of the faculty responsible for contemplating wonder and mystery.[28] Although the Phrenological Society had expressly egalitarian aims, Coates celebrated the creative potential of the properly cultivated mind by holding up a transcendent ideal that men could aspire to. This tremendous optimism about the potential and plasticity of the creative faculties would contribute to Coates's medical theories on the effects of alcohol abuse and his temperance activism in the late 1820s.

In addition to their central importance to the Phrenological Society, Coates, Bell, Condie, and Horner linked their medical careers to the new physiological systems of medicine being developed in France, especially the theories of François Broussais. In 1826 Bell helped translate and publish Broussais's *Treatise on Physiology Applied to Pathology.*[29] In 1829 Horner wrote the first American work on pathological anatomy, which followed the teachings of Broussais very closely, and Coates taught private courses in physiology and lectured on the new medicine to the managers of the Pennsylvania Hospital.[30] The theories of Broussais profoundly influenced physicians' ideas about the effects of alcohol abuse on the body and mind and spurred the new emphasis in medical education on postmortem dissection. While French theories of medicine stimulated important changes in medical education and practice in Philadelphia in the 1820s, advocates of physiology cast the new medicine as both a medical and a social reform movement.[31]

Physiology revolutionized how doctors constructed authority, both within the profession and in relation to potential patients. The application of French physiology rested on direct observation, open for all to see. Especially through postmortem dissection, physicians now justified their truth claims and treatment strategies by empirically demonstrating the seats of disease in the internal organs. French physiology enabled American physicians to declare an allegiance to visible truth, which took on "powerful symbolic significance" in a society "paranoid about being tricked by such archetypal counterfeits as the painted woman and the confidence man."[32] Physiology also worked to guarantee university-trained physicians exclusive medical authority, since they alone had the knowledge gained through postmortem dissection, and promised a new relationship between the profession and the public.

Coates, Bell, Condie, and Horner, like other ambitious young doctors, argued that phrenology and physiology offered a new direction for American medicine, and they sought to lead these reform efforts. Coates's career particularly illustrates how these efforts reflected his professional ambitions and the rigorous demands of gaining membership in the upper echelons of the Philadelphia medical establishment. Embarking on his career, Coates enjoyed advantages that accrued only to the most privileged. His family was wealthy, and Benjamin's father, Samuel Coates, was a merchant and philanthropist who had served as a manager at the Pennsylvania Hospital from 1785 to 1825. Samuel apprenticed both his sons at the hospital. Benjamin's younger brother Reynell received his MD from the University of Pennsylvania in 1823. Benjamin and Reynell inherited their father's extensive business holdings, which Benjamin managed after his father's death in 1830.[33]

Constructing a private practice was difficult, and even the most successful physicians led hectic and exhausting professional lives.[34] But medical professors were far and away the best known and wealthiest physicians in Philadelphia, the capital of American medicine. Lecturing on physiology and promoting phrenology was part of Benjamin's broader campaign to win a place in that elite brotherhood. In the mid-1820s he became an independent instructor and found that competition for students was fierce.[35] Hoping to find strength in numbers, Coates and Condie founded the Medical Lyceum of Philadelphia, a confederation of private teachers working together to provide a more complete and thorough medical curriculum. Though the Lyceum lasted only a couple of years, Coates continued as an independent instructor in the city.[36] Along with Coates and Condie, Bell also tried to establish himself in teaching. Although he did not teach at the Lyceum, he did win an appointment at the prestigious Philadelphia Medical Institute, a summer medical institute established by professors from the University of Pennsylvania.[37]

Coates's address opening the winter session of the Lyceum in 1825 suggests the tremendous energy and excitement generated by the expansion of education, research, and publication. He celebrated the rapid developments within the American medical profession and likened them to the remarkable technological innovations transforming all aspects of American life. "In the progress of time," he told his students, "and in an era of such wonderful changes as are constantly transpiring around us, it was not to be supposed that the establishments for medical instruction would stand still." Coates celebrated America's industrial achievements while extolling the virtues of democracy and territorial expansion:

> While new facilities for the advantageous employment of industry are continually starting into existence, while new modes of communication between distant portions of the world . . . are gradually extending themselves around the globe, while canals are succeeding canals and rivers after rivers are subjected to human control, while new empires have sprung up, opening an immense commerce, independent & fearless of the despotisms of continental Europe, and rendering America, at length, truly worthy of the long-appropriated title of a "new world," . . . what character would our physicians have deserved, if, amid this mighty, moving mass, their profession had alone been stationary?[38]

Coates's professional aspirations, patriotism, and sense of social responsibility were inseparable. His lofty nationalist optimism also typified the claims being made by the standard-bearers of the medical profession. Coates and other elite physicians shared a grandiose vision of their profession's mission to serve a benevolent, democratic American empire.

As an extension of his work to advance himself as an instructor, Coates embarked on a costly and risky effort to promote French physiology within the medical profession by founding the *North American Medical and Surgical Journal* in 1826. This new journal entered a crowded field.[39] At least one of Coates's friends questioned whether it could be successful given the number of journals being published and the high failure rate.[40] But editing a medical journal gave Coates distinct professional advantages, and his family wealth supported the endeavor. By founding his own journal,

Coates claimed a central place in a network of correspondence with other medical men and had an outlet for his own essays. The journal made him a more attractive teacher, both because it lent him prestige and because young physicians in training were eager to publish. The journal also placed Coates in the vanguard of those promoting theories of physiological medicine.

Advancing medical knowledge through research and publication was becoming vital to accruing a reputation, attracting students, and winning a place among Philadelphia's medical elite. Coates contributed articles on asthma and inflammation of the larger arteries to the *North American Medical and Surgical Journal*, but his major piece was on delirium tremens. He published it in July 1827, the same summer the temperance movement was formally launched in Philadelphia. He wrote in part out of his contempt for the delirium tremens treatment proposed by Joseph Klapp, whose radical emetic cure had dominated local discussion of the disease in the early 1820s.[41] Coates believed Klapp's cure was poisonous and argued for using large doses of opium.[42] Klapp and Coates continued their contentious public debate over delirium tremens treatment for at least ten years.[43]

Delirium tremens certainly fit Coates's intellectual interests, given its association with a diseased imagination. As demonstrated by his ideality lecture, Coates was a romantic, deeply interested in the workings of imagination and fancy. He actively participated in Philadelphia's literary circles, contributing poetry and essays to magazines anonymously.[44] He also harbored a deep interest in hallucinations, ghosts, and the supernatural. During the 1820s and 1830s, he periodically checked out books on apparitions from the Pennsylvania Hospital library.[45] His interest in the nature of insanity was a family affair: his father had kept a handwritten diary describing the fanciful delusions of lunatics in the insane asylum at the Pennsylvania Hospital. One inmate, Richard Nesbitt, had particularly fascinated Samuel Coates.[46] Nesbitt was a long-term inmate known for his brilliant poetry and imaginative paintings, one of which he gave to Benjamin Rush's son James.[47] Late in his life, Samuel bequeathed his record book to his son Benjamin.[48] For the Coates family, the diseased imagination continued to be irresistible.

Coates's diverse professional efforts came to fruition in 1828 when the managers of the Pennsylvania Hospital appointed him to a high-profile position as one of the hospital's attending physicians. He oversaw the lunatics at the hospital until the new asylum opened in 1841. No doubt his father had something to do with the appointment; James Rush, Benjamin's son, was also an attending physician at the hospital, so family ties were clearly meaningful to the hospital's managers. But Coates's article on delirium tremens no doubt helped convince them of his qualifications.

Physician Activists

When the first temperance societies were formed in Philadelphia in 1827, intellectual movements, professional pressures, and competitive dynamics spurred the medical community to action. Meigs, Coates, Bell, and Condie were all leaders of the statewide Pennsylvania Society for Discouraging the Use of Ardent Spirits. The four men also participated in local societies as activists, writers, and speakers. In 1828, for instance, Meigs, Bell, and Condie worked as managers of the Young Men's Association of Philadelphia for the Suppression of Intemperance.[49] Three physicians served on that association's seven-person executive committee, and seven of the twenty-one members of its board were also doctors.[50] Temperance activism was an enduring tradition among Philadelphia physicians: Benjamin Rush had been training doctors to be temperance activists since the 1790s. But in the 1820s, temperance also fit with the new ideas and impulses moving through the profession. Promoting physiology, phrenology, and temperance served the professional interests of some of the city's most ambitious physicians.

Temperance had broad support within the American medical community. Eighty Philadelphia physicians signed a petition backing the temperance movement, which stated unequivocally that heavy drinking represented "a frequent cause of disease and death."[51] As one put it, the call to temperance activism demanded that doctors live up to their unique position as "guardians of the public health."[52] Doctors throughout the country issued similar appeals. In 1829, noting that "ten Medical Societies, in different parts of the United States," had passed resolutions in support of temperance, the New York City Temperance Society published an appeal to the city's physicians. "Their peculiar station," the pamphlet asserted, gives doctors "facilities for the counteraction of the evil which no other persons possess."[53]

Through the early 1830s, physicians, noted evangelicals, and wealthy businessmen dominated the leadership of temperance societies. The Pennsylvania Society's work fell into two related categories, both shaped by the class interests of its founders. First, it held up to public view the social consequences of the widespread use of ardent spirits. Activists quantified the economic costs of poverty, crime, and disease created by the sale of ardent spirits, demonstrating the burden placed on taxpayers, who funded the city's public agencies. Second, the society hoped to persuade respectable and temperate young men that ardent spirits threatened their health and well-being. Like phrenology and physiology, temperance societies' core message linked physical health with the promise of social achievement. The founding of the Pennsylvania Society was motivated by "the desolating effects produced upon society, by this insidious destroyer." Leading the society, elite citizens sought to publicize information on "the dreadful consequences of indulgence" and promote "the benefit resulting by abstaining from . . . this pernicious drug." Toward the ultimate goal of abolishing ardent spirits, the society hoped to remold the opinions, habits, and fashions governing drink: "Reflecting upon that principle in our nature, which causes men to yield to the force of example, and the influence of fashion, they have resorted to two most powerful weapons, *moral suasion* and *public opinion*, to combat this monster."[54]

The Pennsylvania Society for Discouraging the Use of Ardent Spirits also moved to promote temperance as decent masculine behavior.[55] In Philadelphia the society sent an urgent appeal for the aid of the city's ministers and employed a full-time lecturer to speak in churches and other venues throughout the city and state. The society acted as an umbrella organization, coordinating the work of local societies and encouraging new chapters. The Pennsylvania temperance society oversaw, for instance, the formation of the Young Men's Association of the City and County of Philadelphia for the Suppression of Intemperance. This group advanced the temperance cause by the "influence of moral example, in abstaining on all occasions from the use of ardent spirits, and by appeal to the reason, hearts, and consciences of men, endeavour to dissuade them from its use in like manner."[56] By 1832, temperance advocates claimed that the city of Philadelphia boasted over thirty temperance societies with a total membership of 4,500.[57]

Despite its egalitarian rhetoric, the efforts of the Pennsylvania Society for Discouraging the Use of Ardent Spirits betrayed growing social divisions, especially on the subjects of urban poverty and the habits of workers. Activists reprinted the findings of reports, especially the 1817 Report of the Pennsylvania Society for the Promotion of Public Economy, arguing that Philadelphia's poverty derived overwhelmingly from intemperate habits.[58] At the same time, the Pennsylvania Temperance Society strongly doubted that it was possible to reform the drinking habits of the impoverished. The report by the Philadelphia Medical Society argued that "the instances of recovery from habits of intoxication, though such sometimes occur, are unhappily so rare as to leave but little encouragement for efforts in these quarters."[59] The New York Society was not calling on physicians to reform drunkards either; rather, "Our appeal is to the temperate, the only class of society with whom we may hope to succeed."[60] While highlighting the consequences for intemperate paupers and criminals, these physician activists worked only to persuade respectable and worthwhile individuals to pursue a life of self-restraint so as to avoid becoming degenerate paupers themselves. Only proper habits could ensure that young men did not fall into poverty and depravity. The intemperate poor and criminal were irredeemable.

Further, while temperance leaders came largely from the city's elite, including wealthy industrialists, they focused their reform efforts on young mechanics and journeymen. The Pennsylvania Society sent hired lecturers primarily to the emerging industrial districts of the Northern Liberties, Kensington, and Southwark, although only 20 percent of the leaders of temperance societies lived in these outlying areas.[61] In these neighborhoods, activists worked to change the custom of drinking on the shop floor, causing resentment among those who actually worked there. These class tensions occasionally were expressed in violence. Angry mobs attacked temperance meetings several times during the 1830s. Activists blamed the violence on the evils of drink and the owners of grog shops.[62]

The Pennsylvania Society for Discouraging the Use of Ardent Spirits framed the problem of intemperance in many ways that revealed the concerns of its elite leaders. It commissioned reports to determine the number of taverns in the city's poorest neighbourhoods and investigated state laws on drinking and gambling. The society also researched the feasibility of encouraging the cultivation of grapes for wine as an alternative to spirits. In the 1830s, temperance advocates used these many reports as the bedrock of their appeals. Reformers grew increasingly sophisticated at presenting statistics, such as the number of gallons of liquor consumed annually in the United States, the number of criminals imprisoned for offenses committed under the influence of alcohol, and the number of paupers rendered dependent and confined to the almshouse because of intemperance. By 1833 temperance advocates broadened their claims to speak to the national importance of the temperance campaign, calculating that the financial burden to the nation created by the unnecessary consumption of ardent spirits totaled $94,525,000, although they cautioned that the real costs were far higher: "In this estimate, no account is taken of the loss of the labour of the paupers, prisoners confined for debt, nor of the costs of litigation created or excited by the use of ardent spirits, nor the salaries of judges, the expenses of juries, nor of the fees of counsel."[63]

Reformers' appeals thus targeted the concerns of persons of some means and property. Ridding society of intemperate drinking would lift a heavy financial burden from middle-class taxpayers while creating a more efficient and industrious workforce, which they believed meant economic benefits for all citizens.

Writing on behalf of the Pennsylvania Society for Discouraging the Use of Ardent Spirits, temperance physicians similarly betrayed the class interests of the society's founders and leaders. Whether by design or not, Coates's essay on delirium tremens made him Philadelphia's leading authority on the problem of alcohol abuse at the moment when temperance organizations appeared in the city and state. In 1829, when the Philadelphia Medical Society appointed a committee of five physicians to write a report on the health effects of intemperance, they chose Coates as the principal author.[64] The city's magazines and newspapers widely reprinted the report, which was the first major medical publication on temperance since Benjamin Rush died in 1813.[65] The report was a departure from Rush's famous pamphlet *An Inquiry into the Effects of Ardent Spirits upon the Human Body and Mind*. In part this departure reflected intellectual developments within the medical profession. Coates took aim at what he characterized as antiquated systems of medicine. He particularly criticized older theories that called for treating disease by heavy use of powerful stimulants such as liquor. Rather, the report stated, "What has been called the *physiological medicine*, goes still farther than any former doctrine to discourage the unnecessary employment of spirituous liquors."[66] Coates concluded that "under ordinary circumstances, ardent spirits, in any quantity, whether great or small, are injurious to the health of the system. . . . Pure water is confessedly the most natural and most proper drink of man." Coates used physiological principles to bolster the temperance cause, but he also used the strong popular sentiment for temperance reform to further his commitment to elevating the principles of physiological medicine within the American medical profession.

Coates's report also demonstrates a turn away from Rush's central concern with preserving republican virtue and toward a medicine shaped by the economic and social concerns of the new middle class. Coates shared Rush's focus on liquor's contributions to disease, criminality, and death, but his essay's central concerns differed. Rush had emphasized that drunkenness threatened the moral faculty and thus the survival of republican government; Coates emphasized liquor's role in producing widespread poverty and loss of respectability. Rush had hoped to shape the temperate behavioral norms he saw as essential for a virtuous electorate; Coates's report acknowledged and legitimated the growing disparities of wealth and class emerging in 1820s Philadelphia.

In summarizing society's stake in discouraging the use of ardent spirits, the report did not once mention government institutions. Betraying the ideological nature that physiology took on in the 1820s, Coates equated physical health with social status and rectitude, writing that liquor threatened the destruction of "health, strength, riches and respectability, and . . . the future misery of an immortal soul."[67] While liquor posed a dire threat to human health, Coates further cited the report on poverty published in 1817 by the Society for the Promotion of Public Economy as evidence that heavy drinking was the primary cause of poverty in Philadelphia. He reinforced these attitudes by offering his professional experience: "We have frequent and melancholy opportunities of witnessing, in the abodes of the unfortunate, the manner in which pecuniary difficulties are generated; and we believe it is the universal sentiment of those who possess such means of information, that the greater portion of the existing distress in this country, is the result of the employment of ardent spirits."[68]

If poverty derived primarily from intemperance, Coates saw no hope for reforming inveterate drinkers.[69] Whereas Rush had sought to enlist the medical profession in shaping republican citizens, by the late 1820s the prevailing view among Philadelphia physicians was that habitual intemperance was not subject to medical treatment.[70] Coates's report reinforced the linkage of drinking with poverty. He made it clear that physicians should concentrate solely on convincing the respectable and already temperate of the dire health dangers of casual drinking.

The publications and actions of the Pennsylvania Society for Discouraging the Use of Ardent Spirits provoked anger from some quarters. Even the meaning of "temperance" became highly contested. Bound up in republican ideals celebrating personal independence and virtuous behavior, the notion that temperance in eating, drinking, and smoking constituted manly self-restraint resonated, for instance, with workers as much as with prominent middle-class citizens.[71] But what constituted temperate behavior was very much up for debate. Temperance was a common subject in the *Mechanic's Free Press*, a publication closely associated with the Philadelphia Workingmen's Party. First published in 1828, the newspaper aimed to unite "working people in one firm body, for the maintenance of their rights, the promotion of their interests, and the obtaining of that control in the making and administering the laws which their numbers, usefulness, and intelligence entitle them to."[72] Some correspondents' letters published in the paper portrayed excessive drunkenness among young men as a failure by greedy employers to take proper responsibility for their apprentices' moral conduct.[73]

More often, however, short stories and poems in the *Mechanic's Free Press* portrayed drunkenness as a violation of the father's masculine duty to protect and provide for the family and reinforced the linkage between poverty and drunken depravity. Printed in 1828, "The Drunkard" was a typical temperance poem:

> I saw him, 'twas at dawn of day
> Before an Ale House door;
> His eyes were sunk, his lips were parch'd
> I view'd him o'er and o'er
> His infant boy clung to his side, and lisping to him said,
> "Come father—mother's sick at home,
> And sister cries for bread."[74]

Another newspaper item warned that a drunkard who was "a married man a father of sons and daughters, all smiling, or willing to smile, round his board . . . deserves that death should come stealthy in, once a month, like an unseen tiger at midnight and carry them all off, one by one to his den the grave." The author invoked the seventeenth-century authority Robert Burton in warning, "'If a drunken man,' quoth old Burton, in his *Anatomy of Melancholy*, 'gets a child, it will never likely have a good brain.'"[75] As an ideal, temperance had broad appeal as responsible masculine behavior.

Nevertheless, many contributors to the *Mechanic's Free Press* viewed the Pennsylvania Society for Discouraging the Use of Ardent Spirits with deep suspicion. In 1828 the correspondent "Equity" called it "laudable" that "there is an effort making [*sic*] by a certain class of citizens to promote temperance." Equity deeply resented temperance advocates who portrayed alcohol abuse as exclusively a workingman's problem. The rich have the means to hide their drunkenness, he asserted: "The most foul and beastly intemperance is daily practiced by persons who have it in their power to keep at such times from public view." True temperate behavior, he argued, would mean not just abstaining from drink, but also showing restraint "in exacting severe toil and labour from those whom we employ. In our pursuits of wealth; in our gratification of either natural or artificial taste, and above all in our deportment, that it savor not with self sufficiency and exaltation . . . while we see others round us destitute of the common comforts of life."

Equity saved his harshest condemnation for wealthy and religious temperance advocates who stigmatized the poor as intemperate:

> It is evident at the present day that there is every effort making [*sic*] to impress on the minds of the poor, the necessity of contentment under the most severe privations, by those who only wish more effectually to rivet the galling yoke of oppression by superstition and priest craft on their necks, and who would not lift so much as the weight of their little finger to ameliorate their condition.[76]

For contributors to the *Mechanic's Free Press*, the temperance cause was not itself divisive. Rather, they saw the actions taken by wealthy activists in the name of temperance as hypocritical, stigmatizing workingmen and the poor.

By the 1840s, the character of temperance activism and the social makeup of activists changed considerably. In Philadelphia, middle-class dominance of temperance organizations began to erode as groups proliferated. After 1835, journeymen, skilled workers, and other working people combined the anti-liquor campaign with communal efforts to protect their families from the uncertain economy.[77] Journeymen mechanics often led these meetings.[78] Especially during the eight-year depression that followed the Panic of 1837, involvement in temperance beneficial societies became a coping strategy for struggling workingmen and their families.[79] The leaders that emerged brought a new sensibility to temperance activism. Noted lecturer Lewis Levine was "crude, vulgar, and something of a charlatan with a flair for demagoguery and a hunger for political office."[80] Trained as a lawyer, Levine had a rough style that appealed to men across the social spectrum—workingmen as well as members of the protestant middle class. In the 1840s, he moved on from temperance to nativist politics. In 1844 he helped incite anti-Catholic riots in Philadelphia's Irish neighborhoods. In the aftermath, he emerged as a nationally known figure in the nativist politics of the 1840s.[81]

In 1840 the founding in Baltimore of the Washington Temperance Society dramatically broadened the appeal of temperance among working people nationwide. The Washingtonians sought to reach out to confirmed drunkards, reform their habits, and give them protection and support in living a sober life.[82] Largely secular, the Washingtonians structured their meetings around recovered drunkards' highly emotional accounts of the destruction alcohol had wreaked on their lives and families.[83] The movement was a national sensation: one estimate is that 600,000 people joined Washingtonian societies between 1840 and 1845.[84] At popular "conversational meetings," as one newspaper announcement described it, former drinkers "relate their *experiences* and show up the *morality, benefit,* and *happiness* resulting from the use of *distilled damnation.*" This notice pointedly observed that "Tavern Keepers and *grog Shop Keepers,* and *Drinkers of Liquors* are invited to attend."[85] In Philadelphia, newspaper advertisements demonstrate that temperance societies of all kinds, not just those inspired by the Washingtonian model, proliferated in early 1841. The *Public Ledger* reported that in the first three months of that year, 4,300 people in the city had newly joined temperance societies, bringing the total number of its citizens involved in such organizations to 17,000.[86]

Even as the temperance movement and its leaders became more socially diverse, however, physicians continued to be important in leadership and in shaping temperance literature. In 1837 Condie was a founder and the first president of the Temperance Beneficial Association in Southwark, one of the city's poorest neighborhoods. These temperance beneficial societies emphasized the plight of families, encouraging the founding of women's chapters to supplement all-male meetings.[87] The association's Southwark Branch No. 1 formed a model for others, and by 1841 at least twenty branches of the Temperance Beneficial Association held frequent meetings throughout the city.[88]

Condie's leadership suggests that physicians' appeal to health and well-being enabled them to rise above the accusations of class interest or priestcraft that some labor activists leveled at the early temperance societies. A resident of Southwark, Condie led elite philanthropic organizations as well. In 1840 he was one of six doctors who led the Philadelphia Temperance and Benevolent Association. As a central part of its work, the group divided the city into nine districts and assigned a physician to each. These doctors ministered to the illnesses of chosen poor families.[89] The association was as much a paternalistic effort by city elite to relieve poverty as it was a temperance organization. In addition to sponsoring temperance meetings and spreading temperance literature, the group sought out respectable poor families to receive food and other assistance. The founders carefully distinguished their ministrations from "indiscriminate charity." By requiring a temperance pledge, they hoped to eliminate the cause of poverty even as they ministered to the poor: "Scarcely a suffering invades the domestic hearth—scarcely a vice deforms the human character—scarcely a crime appears upon the criminal calendar that does not claim relationship with, or acknowledge as its parent, the alcoholic poison, and among the poor, the suffering poor, are found the greater number of its victims."[90]

But the association was not dedicated to reforming drunkards. It required that a chapter of the Temperance Beneficial Association or some other reputable source recommend respectable poor families, and all families had to sign the temperance pledge before receiving aid. As a physician, Condie was thus able to participate in mutual benefit societies while at the same time leading conservative temperance groups dominated by wealthy citizens deeply suspicious of the moral habits of the poor and unfortunate.

A Temperate Profession

Within the medical profession, support for temperance extended well beyond ambitious doctors like Condie, Bell, and Coates. For ordinary practicing physicians, especially those just embarking on their careers, temperance activism helped with the peculiar economic pressures they faced. Medical students were keenly aware that building a private practice was a challenge.[91] Before the Civil War there were no fixed career paths for these young men. They had to strike out on their own. In the 1820s and 1830s, professional organizations were weak and offered little help. Licensing laws were ineffective, and in the 1830s most were abolished, meaning anyone could practice medicine, regardless of training. The attraction of a university degree was its prestige and the confidence it inspired in prospective patients. Many students attended lectures in the city for only a year, however, and moved on to practice medicine without graduating. Many failed to finish the degree because it was difficult to recoup the costs of a lengthy education. The income from private practice was often small, and many doctors supplemented their income with other work, such as farming. In rural areas where currency was in short supply, collecting payment for medical services was difficult.[92]

Even considerable wealth and education did not ensure professional success. Reynell Coates had the same advantages of training, wealth, and connections as his older brother Benjamin, but when it came time to embark on his own career, he became discouraged by the intense competition in Philadelphia, as many doctors did. He left to build a practice in western Pennsylvania but found the endeavor very difficult. He struggled to collect from his cash-poor patients and complained to his brother about the isolation of rural life. Far from the wealth and comforts he had enjoyed in his father's home, he wrote, "I cannot long endure the kind of life I now lead. I shall lose my senses if I continue to be deprived of all society here, and as to my returning to Philadelphia, I think there is as strong a prospect of my arriving at the far side of the moon."[93] Despite his inherited wealth, prestigious training, and distinguished family name, Reynell's finances remained chaotic for decades, and he went bankrupt in 1844.[94]

With its rapid growth in the 1810s and 1820s, the profession needed a larger market of paying clients. Since the colonial era, however, university-trained physicians had been in short supply, and American families had learned to be self-reliant about medical care. In the 1830s most Americans lived in rural areas, far from the hospitals and almshouse wards that provided emergency medical care in large cities. They relied on well-established traditions of vernacular medicine and on family members, midwives, local healers, good Samaritans, and others who responded to need.[95] They were aided by an abundance of popular guides to domestic practice, including reprints of William Buchan's *Domestic Medicine*, first published in 1769, and *Gunn's Domestic Medicine, or Poor Man's Friend*, first published in 1830 by Tennessee physician John C. Gunn.[96] Home medical guides were supplemented by an array of patent medicines sold by peddlers, pharmacists, and grocery stores that promised to treat numerous afflictions. Early American newspapers were filled with advertisements for popular medical texts and products.[97]

University-trained physicians also found they had to defend themselves against claims made by practitioners of rival systems of medicine. Mesmerists, magnetic healers, homeopaths, and others provided stiff competition to orthodox physicians throughout the antebellum period.[98] The largest organized health movement was the national network of Thomsonians. Published in 1822 by Samuel Thomson, who had no formal medical training, the *New Guide to Health* outlined a system of botanical medicine, which Thomson cast as a commonsense alternative to arcane professional medicine. Spreading from rural New England into western New York and Ohio, local Thomsonian practitioners organized themselves into a network of societies connected by a nationally circulating journal.[99] A central part of the popularity of health providers like the Thomsonians was their anti-elitist assertion that medicine should be demystified and open to everyone. They accused university-trained physicians of defrauding their patients with expensive, useless, or damaging therapies.[100] The critique fell heavily on traditional therapies that were highly intrusive, such as excessive bloodletting and vomiting, or the heavy use of opium or alcohol. In the 1830s, Thomsonian and other unorthodox healers succeeded in pressuring state legislators to abolish many ineffective licensing laws, rendering the medical marketplace even more wide open.

Without medical licensing, powerful professional associations, or large institutions, orthodox doctors had to win patients through their medical practice. Inspiring confidence was imperative.[101] Writing to Benjamin Coates from Pine Woods, Louisiana, one physician recounted travels in search of a private practice that took him from Cincinnati to New Orleans via St. Louis and Charleston. "Where is there not competition?" he asked. "Money making is the spirit of the times, whether by merit or by imposition." Aspiring doctors had to consider two major factors when evaluating the opportunities in a community: "First, the number of physicians," and "2nd the degree of confidence reposed in them." The first, he continued, was not nearly as important as the second. Opportunity arose in communities where patients did not trust their present doctors. Patients' confidence, he wrote, is "of the utmost importance to a young man & his immediate success must be governed by it."[102]

These pressures made professional demeanor crucial for young physicians.[103] Professors urged their students to cultivate a strong moral character, to devote their lives to disciplined study, and above all to avoid drinking.[104] These appeals were not new; Rush had often implored his students to remain temperate. But the temperance movement of the late 1820s invigorated these calls within the profession. The Philadelphia Medical Society called on physicians to live up to their unique role as guardians of public health "by refraining from the intemperate use of alcoholic liquors."[105]

For doctors seeking to build a private practice, adhering to this maxim was sometimes difficult. Temperate behavior could be important for physicians whose prospects depended on inspiring confidence in prospective patients, but inspiring confidence above all meant reflecting the expectations and norms of the community where they worked.[106] A leading physician in the West, Daniel Drake noted that customary hospitality rendered doctors particularly vulnerable to intemperance. "In a country where ardent spirits are a constant offering of hospitality," doctors who regularly called on patients in remote areas were prone "to drink too much." He recommended that "no physician should allow himself to use ardent spirits, until he has passed his fortieth year," when he was old enough that he was unlikely to fall into intemperate habits. Speaking in 1828, Drake believed this was a particular challenge for rural doctors: "In our cities, the number of intemperate physicians is fewer than it formerly was, but in the villages and the country, they are still so numerous, as to bring on the profession discredit."[107] Physicians had to negotiate a social landscape in which drinking had become increasingly controversial while offering liquor remained a prevalent form of hospitality.

Supporting the temperance cause thus represented a declaration of allegiance to the values, concerns, and codes of behavior central to the new urban middle class. Above all, it offered a way for physicians to link strong moral character and social responsibility with their medical practice. By the early 1830s, medical schools nationwide began forming their own temperance societies. The Physiological Temperance Society of the Medical Institute in Louisville urged practicing physicians to cooperate with temperance activists because of "their superior knowledge of the subject and the confidence which the community at large repose in them on a subject so strictly professional." Further, by refraining from drinking themselves, doctors lent "authority to their precepts and efficacy to their exhortation."[108] These concerns were also at the heart of the temperance societies founded for medical students at the New York College of Surgeons and Physicians and the University of Pennsylvania.[109] In his 1833 address to the University of Pennsylvania society, Bell called temperance "the promoter of health, comfort, intellectual excellence, and of all good and noble works." Significantly, he emphasized that modeling temperance not only would make them better doctors but would also improve their social standing. "The sphere of your future usefulness will be enlarged in a singular and pleasing manner by your adding personal example to doctrine," Bell urged, "and thus encouraging others to adopt a similar course."[110] For university-trained physicians, temperance offered a way to present orthodox medicine as being in the service of the greater public good, at least as defined by socially respectable patients.

Drawing on the principles of physiology, Bell implored these young students to recognize that temperance was above all good medicine, and that identifying with the cause would demonstrate the efficacy of university-trained physicians compared with unorthodox providers.[111] Physiologists, especially Broussais, believed that disease derived from repeated morbid stimulation of the stomach and lungs. In his address to the Medical Students' Temperance Society, Bell made it clear how diet affected physiology:

> No diffusible stimulus, no vegetable or mineral tonic, nor even a mild bitter, such as chamomile tea for example, can be continued to be used habitually, without injury resulting,—first to digestion,—afterwards to the nervous and blood-vessel systems. With how little show, then, of reason, not to say of physiology, can we admit of the regular daily use of . . . ardent spirit?[112]

Promoting temperance among patients not only was good for health, it was also a way for these would-be physicians to demonstrate the value of their university training. Temperance, Bell said, "is the cause of rational medicine," presumably in contrast to irrational systems of medicine.

At the heart of the temperance impulse within American medicine was thus an effort to demonstrate that the principles of orthodox medicine were in the interest of ordinary people. This impulse extended beyond a renewed emphasis on temperate comportment and social usefulness. Leading physicians capitalized on the swelling popularity of the temperance movement to promote orthodox medical knowledge and practice as transparent, meaningful, and relevant. The goal of demystifying medicine lay behind efforts by Bell and Condie to publish a journal aimed at a popular audience. Issued biweekly from 1829 to 1833, the *Journal of Health* focused heavily on the implications of the new physiological medicine. The journal was part of the founders' temperance activism. Indeed, Bell wrote the Pennsylvania Temperance Society's 1831 annual report, and the report was printed and distributed through the offices of the *Journal of Health*. Through the journal, Bell and Condie sought to capitalize on the growing public support for temperance by reshaping popular ideas about health to conform to new physiological principles, and to answer popular criticisms and suspicions of orthodox medical practice. In their first issue, the editors announced that "deeply impressed with a belief, that mankind might be saved a large amount of suffering and disease, by a suitable knowledge of the natural laws to which the human frame is subjected, they propose laying down plain precepts, in easy style and familiar language, for the regulation of all the physical agents necessary to health." In particular, they said, the journal would emphasize "the value of dietetic rules . . . and the blessings of temperance . . . with emphasis proportionate to their high importance and deplorable neglect."[113]

True to its stated goals, the *Journal of Health* exerted broad influence on public opinion, broader in some ways than the influence achieved by the temperance organizations of the late 1820s. The journal was regularly excerpted and cited in both medical and nonmedical publications.[114] Despite suspicion of the actions taken by the Pennsylvania Society for Discouraging the Use of Ardent Spirits—an organization with which Bell and Condie were closely identified—the *Mechanic's Free Press* informed its readers that the *Journal of Health* was an admirable enterprise. The newspaper's editors praised the journal as "a valuable work and calculated to do much good: it is decidedly opposed to quackery, as well as to the strong predilection which medical men generally evince, to keep us in ignorance as to the causes of the various diseases to which the human system is liable and the means by which they may be eradicated."[115] The journal's appeal to health transcended the class tensions that other forms of temperance activism sometimes engendered.

Despite the endorsement of the *Mechanic's Free Press*, the *Journal of Health* strongly identified with the concerns of the conservative temperance associations of the late 1820s and early 1830s. This identification was made explicit in an article titled "The Middle Classes." "Among what class in society . . . are we to seek for the greatest amount of health?" the journal asked. "The rural classes—the decent citizens—people possessed of education and employment, but neither over-refined, nor over-worked—the farmer and moderate proprietor—the man of action and of enterprise . . . it is in their ranks that the medical philosopher finds health and happiness best established."[116] Even if the appeal to healthy moderation, exercise, and self-restraint reached across class lines, the journal championed habits and values that were unattainable by the city's workers and laboring poor.

The Eclectic Medical Marketplace

The powerful appeal of a medicine that linked the rhetoric of temperance, industry, physical health, and social well-being turned out to be something of a two-edged sword. Excerpts from the *Journal of Health* reached a wide audience, raising awareness of physiological and phrenological principles. The journal also contributed to the diversification of unorthodox medical providers. By the 1830s and 1840s, growing public awareness of anatomy and physiology, and the idea that these principles could be applied to social problems, fueled a number of popular health movements, adding to the competition faced by university-trained physicians. Unorthodox medical providers also drew on temperance to represent the efficacy and usefulness of their own systems.

The career of evangelical minister and temperance lecturer Sylvester Graham exemplifies these developments. While employed as a lecturer for the Pennsylvania Society for Discouraging the Use of Ardent Spirits, Graham took up the concerns aired in the *Journal of Health* and built a career as a zealous health reformer. He formed his own physiological system based on the theories of the French physicians François Broussais and Xavier Bichat. Taking the health consequences of liquor as his starting point, Graham soon began campaigning against masturbation, refined flour, spicy foods, and all other forms of what he saw as morbid stimulation of the body.

He published a total system of personal health and hygiene, teaching followers to follow a carefully controlled daily regimen that included a bland vegetarian diet, cold water, and exercise at specific times of day.[117] Early in his career, physicians praised him as a useful advocate for health, but some later denounced him as a charlatan in major journals such as the *Boston Medical and Surgical Journal*.[118]

Graham's decline in popularity in the early 1840s did nothing to stem the tide of popular health lecturers, journals, and books. Grahamite journals continued to disseminate his physiological principles and encouraged interest in other alternative systems of medicine. Thomsonians, homeopaths, mesmerists, eclectics, and others flourished at midcentury. In particular, physiology, phrenology, and hydropathy—also known as the water cure—became enormously popular in the 1840s and 1850s. Temperance was the sine qua non of all three of these medical-scientific systems of self-improvement.

From its modest beginnings in Philadelphia, by midcentury American phrenology became a part of mainstream popular culture. Promising scientific knowledge that could direct self-improvement, phrenology's appeal especially resonated during periods of economic crisis. The first phrenological society was founded in the wake of the Panic of 1819, and interest in the science blossomed during the bleak years that followed the Panic of 1837. The tour of British phrenologist George Combe from 1838 to 1840 encouraged the formation of local phrenological societies, which soon sprang up in almost every American city.[119] As the science became more popular, the claims made by phrenologists in America and Europe grew more grandiose. The *Phrenological Journal and Magazine of Moral Science* circulated across the Atlantic, and the stated goal of the international phrenological community was to apply "the facts and principles brought to light by phrenological investigations, to the elucidation and improvement of all matters in any way connected with the training and direction of the human mind."[120]

In the wake of Combe's tour, Orson Squire Fowler and Lorenzo Niles Fowler commercialized phrenology.[121] They established permanent offices in Philadelphia, including a lecture room. By 1840 the Fowler brothers' museum had gathered an "immense variety" of skulls and plaster busts of men, women, and children, meant to exemplify the broad range of mental qualities possible in humans. The display included a bust of Dr. Reynell Coates.[122] Their publishing house, Fowler and Wells, printed a flood of inexpensive publications aimed at a popular audience. While grounded in phrenology, the Fowlers published broadly on popular health, incorporating the principles of physiology, anatomy, vegetarianism, and hydropathy. They sold over half a million books and pamphlets by midcentury. Reaching tens of thousands of readers, their *American Phrenological Journal* and *Phrenological Almanac* addressed topics related to health and self-improvement through temperance, diet, education, good habits, industry, and exercise.[123]

In lectures and published writings, O. S. Fowler argued that phrenological science offered powerful evidence of the necessity of temperance: "To the sneering question often put, 'What possible application can your so called science of bumps and sculls have upon temperance,' I reply, that [phrenology] has a great and most happy application to all his duties and relations to himself, his fellow men, and his God." He first published the pamphlet *Temperance Founded on Phrenology and Physiology* in 1841, and it went through twenty-four editions by 1854. In it Fowler developed ten propositions on the relation between drinking alcohol and various states of mind and body. In endorsing Fowler's pamphlet, the *Boston Medical and Surgical Journal* wrote, "If phrenology supplies cogent reasons for living temperate lives, it is turning the science to a practical account at a momentous period."[124] While phrenology may have provided compelling scientific evidence for the importance of abstaining from drink, it was also true that the temperance movement provided a vehicle for promoting phrenological knowledge as attractive and useful to a broad audience.

Temperance and self-regulation were also important to hydropathy. Hydropaths treated disease by applying cold water to various external parts of the body. Knowledge of the medical system spread through the networks forged by promoters of phrenology, physiology, and temperance. Like Graham, hydropaths soon developed a system for total health based on a strict regimen founded on exercise, diet, and abstinence from alcohol and tobacco. Some of the most prominent practitioners of hydropathy came to the science from involvement in temperance societies and Graham's popular health movement.[125] In 1845, for instance, David Campbell, who had previously managed a boardinghouse founded by Graham and governed by Graham's principles, became general manager of the most famous water cure center in the nation at New Lebanon Springs in upstate New York. The best-known hydropathic physician of the late 1840s, Russell Thacher Trall, came to the science after extensive activism in the temperance movement. He had spent years as an editor of a Washingtonian journal in Albany.[126]

The temperance impulse in American medicine thus did nothing to stem the spread of unorthodox systems of medicine. In founding the

Journal of Health, as well as in their other efforts promoting phrenology and physiology, Bell and Condie had helped inspire a new outpouring of popular medical creativity. That these health movements deployed physiological principles further blurred distinctions between university-trained physicians and unorthodox providers. It would be decades after the Civil War before the American medical profession finally quelled this diverse competition.[127]

In the 1840s, the founding of the American Medical Association marked the beginning of greater organization by the medical profession and a more systematic effort to marginalize nontraditional medical providers as quacks. Nevertheless, the AMA was decades away from achieving an effective corporate identity for American doctors. It was only in the late nineteenth century that physicians' career paths began to rely on factors such as licensing, specialties, and large medical institutions. Before the Civil War, the line between orthodox physicians and unorthodox providers was far less clear than it would become in the twentieth century. That competition forced most university-trained physicians to be pragmatic in their efforts to attract paying clients. Medicine was a negotiation between their university training and the expectations, social mores, and beliefs of patients. Physicians based their professional careers on effective practice, reputation, and standing in the community.

By 1850, physician activists nevertheless had been tremendously successful in achieving their stated goals of raising awareness about the importance of temperance and spreading physiological and phrenological literacy. For temperance societies, doctors' involvement helped raise the profile of the cause. Medical literature provided one of the most compelling and broadly influential elements of temperance propaganda. And the blessings of temperance were amplified by the intense competition between orthodox and unorthodox medical practitioners, all vying to win the confidence of patients—especially well-heeled ones.

In the diverse and eclectic medical marketplace, temperance was at the heart of a broad consensus that equated healthy habits with moral behavior, economic success, and social standing. In mid-nineteenth-century America, of course, this consensus did not reflect reality. Quite the contrary. Certainly heavy drinking and disease could lead to poverty in individual cases, but the sharp growth of social inequality that characterized these years derived from sweeping dynamics remaking the American economy and the borders of the United States, not from poor eating and drinking habits. Parallel with the broader temperance movement, American medicine sanctioned the well-being of those who benefited from the economic transformations of the antebellum decades. Acting in their own professional interests, medical providers trumpeted temperance, health, and wealth as physiological principles, true for all men and women, thus providing a biological basis for emerging social disparities and class differences.

FIVE

The Pathology of Intemperance

Beginning in 1841, Dr. Thomas Sewall began using vivid illustrations of dissected human stomachs to illustrate his public lectures on the "pathology of intemperance," which attracted crowds of up to three thousand in the District of Columbia. Sewell used the illustrations to demonstrate the progressive inflammation of tissue caused by habitual drinking. A leading physician, Sewall held a chair in pathology and the practice of medicine at the Columbian College. The goal of his lectures was to further the cause of temperance by presenting empirical evidence of the physiological dangers of drink. The first drawing pictured a healthy stomach, and the last portrayed the ravaged stomach of a man who had died of delirium tremens. They were large—nine times the size of a normal stomach—and colorful (fig. 7). The *Boston Medical and Surgical Journal* reported that "the blood vessels exhibited on the inflamed mucous coat really look as though they would bleed if roughly handled."[1]

For some in Sewall's lay audience, the drawings inspired shock and horror. One temperance lecturer reported that on examining the plates an "unfortunate drunkard" exclaimed, *"They look as I have often felt."*[2] The scientific evidence that Sewall presented so vividly was in keeping with decades of observation beginning with the first description of delirium tremens in 1813. Countless postmortem examinations done by doctors and medical students had yielded a large body of commentary on the appearance of drunkards' internal organs. Trained at the University of Pennsylvania, Sewall's inspiration for producing color illustrations came from Dr. William E. Horner, one of his former professors and arguably the greatest American anatomist of the era. Sewall's stomachs elaborated on a similar illustration in Horner's important 1829 work, *A Treatise on Pathological Anatomy*, and several of Sewall's illustrations had been drawn from anatomical

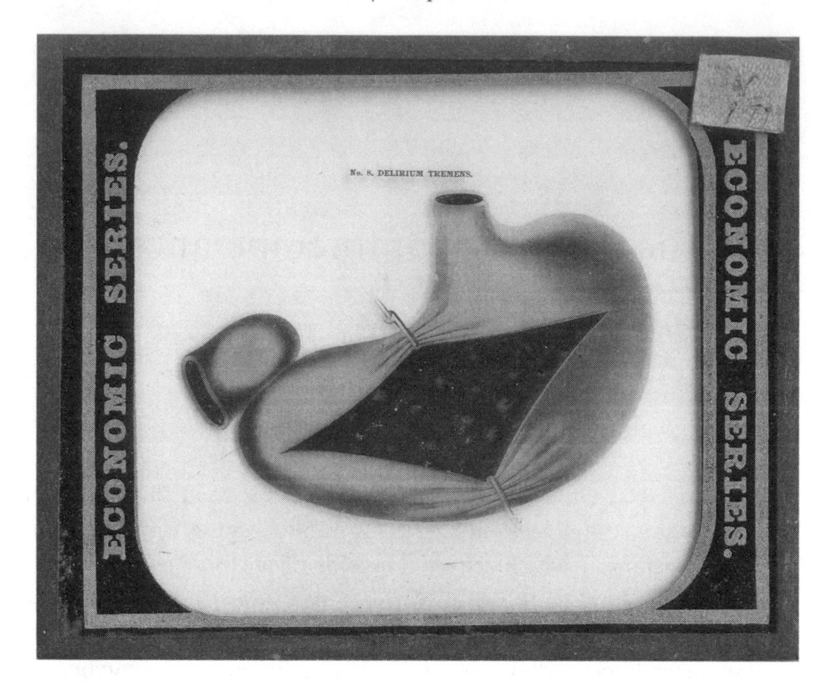

Figure 7. "Delirium Tremens." One in a series of magic-lantern slides copied from Thomas Sewall's "Diagrams of the Human Stomach in Various Conditions." Produced by T. H. McAllister Company, New York, ca. 1850. Collection of the author.

specimens in Horner's collection.[3] Horner himself endorsed the images: "I have looked carefully into your work on the Pathology of the Stomach as influenced by alcoholic drinks and think very highly of the fidelity with which you have portrayed its characters."[4] An impressive array of prominent medical men concurred. Dr. John C. Warren, the Hersey Professor of Anatomy and Surgery at Harvard College, wrote to Sewall, "You could not have resorted to a more forcible and impressive illustration of the fatal effects of this sad and destructive vice."[5]

Sewall's lectures spurred the District of Columbia's temperance reformers to renewed action. Moribund for several years, the Congressional Temperance Society began holding nightly meetings and attracting prominent men to the cause, including Senator Lewis Cass, who would be the Democratic Party's nominee for president in 1848, and Theodore Frelinghuysen, the Whig vice presidential candidate in 1844.[6] The hard-drinking Kentucky congressmen Thomas F. Marshall was Sewall's most notable convert. The nephew of Chief Justice John Marshall, he had attended one of

Sewall's lectures and become convinced that his intemperate habits were life-threatening. He signed the temperance pledge at a meeting held at the Washington Medical College while Sewall looked on. Sewall reported that on signing the pledge, Thomas Marshall "made a most touching speech," and "several other members [of Congress] followed his example." Sewall went on to assert that "Mr. Marshall's step has astonished Congress. There is no man who compares with him in debate."[7] Thomas Marshall's theatrical conversion reinvigorated the Congressional Temperance Society, which lived on for decades.[8]

Sewall's stomachs represent the most graphic example of American physicians' midcentury campaign to publicize new medical theories on the pathology of intemperance. The campaign complemented physicians' concurrent efforts to promote the claim that temperate drinking habits inevitably led to health, economic success, and social well-being. In the 1820s and 1830s, physicians sought to realize the "social usefulness" of the European sciences of physiology and phrenology by establishing them as medical-scientific systems of American self-improvement and social mobility. While working to place medicine in the service of bourgeois uplift, they also cemented in the public mind the physical and psychological horrors that threatened those who deviated from these principles. They warned about many dangers—improper diet, masturbation, opium, illicit sex, and smoking, to name a few, but alcohol was their most pressing worry. Habit-forming, intoxicating, yet still socially acceptable in many households, drinking came to epitomize the opposite of physiological self-discipline. The emergent sciences of temperance and intemperance formed the yin and yang of the new American medicine. Together the pathological anatomy of drunkards and the healthy anatomy of self-regulation formed a new physiology of liberal individualism.

Nothing more clearly demonstrated the importance of proper habits guided by physiological principles than the "catalog of destruction" wreaked by alcohol.[9] Physician activists explained in painstaking detail how intemperate habits caused delirium tremens, varieties of mania, imbecility, and in rare cases spontaneous combustion. Drinkers were vulnerable to life-threatening injury and susceptible to a wide range of diseases including dropsy, apoplexy, cholera, and fever. While many of these claims were familiar, having been made in the eighteenth century and earlier, these nineteenth-century physicians also centered their appeals on a new claim: even casual drinking could quickly become an overwhelming compulsion caused by physical dependence on alcohol.

Given the devastating consequences of intemperance, the news that casual tippling could devolve into depraved compulsion elicited deep public concern. Antebellum physicians successfully spread awareness of this physiological claim, but their appeals had consequences they were not prepared for. As temperance became an ever greater social imperative for the respectable and as the frightening health consequences of intemperance became more widely known, some found long-term abstinence difficult if not impossible. Others came to believe that heavy-drinking family members or other loved ones were subject to a pathological craving. Responding to physicians' claims that drunkards suffered from a disease, some turned to the medical profession for help, but growing demand for treatment extended beyond physicians' ability, or even inclination, to meet.

Historians of the late nineteenth and twentieth centuries have documented the various ways physicians sought to medicalize addiction and thus to enlarge professional authority, create new specialties and career paths, and win state support for building large medical institutions.[10] At this earliest moment in the history of American addiction medicine, however, the medical profession had the opposite impulse. Paradoxically, physicians trumpeted their finding that compulsive drinking was a disease but, at least initially, these same physicians actively discouraged treatments for it. When the doctors were finally spurred to action by patient demand, their efforts remained scattered, experimental, and largely futile. Intellectual developments and professional pressures led them to broadcast the frightening view that the drunkard's imperious craving for drink was an incurable physiological condition. But while they were reluctant to work toward a cure, they nevertheless heightened public fear. In temperance literature, doctors described drunkards' cravings for liquor with fantastic and sensational imagery, often alongside graphic descriptions of alcoholic hallucinations and spontaneous combustion. In the public marketplace these accounts became part of the lexicon of the gothic as writers and lecturers with no medical training imbued the disease with a new emotional power. By the 1840s, lurid first-person narratives detailing struggles with delirium tremens and compulsive drinking became a staple of confessional temperance lectures and fiction. Physicians' efforts to spread awareness about the dangers of alcohol addiction thus inspired a pervasive and prurient fascination with the drunkard's depraved desires and self-destructive compulsion.

An Imperious Craving

Like beauty, what constitutes health and disease is in the eye of the beholder.[11] In late eighteenth-century America, someone who drank eight ounces of liquor over the course of a day would likely have been deemed healthy by most people. In the twenty-first century, many would wonder if that person was alcohol dependent. While theories on the pathology of addiction have varied widely over time and still remain highly controversial, they share the concept that the condition constitutes a self-destructive compulsion. In a 2005 article summarizing addiction research using brain-imaging technology, Wilkie A. Wilson and Cynthia M. Kuhn, both professors of pharmacology at Duke University, offered this definition of addiction:

> Addiction is an overwhelming compulsion. . . . It overrides our ordinary, unaffected judgment. Addiction leads to the continued use of a substance or continuation of a behavior despite extremely negative consequences. An addict will choose the drug or behavior over family, the normal activities of life, employment, and at times even basic survival.[12]

Wilson and Kuhn are among those twenty-first-century researchers who believe that the mechanisms of addiction lie in the brain. Repeated drug use, they theorize, hijacks and warps the powerful mechanisms related to self-preservation. Fundamental processes like finding food to avoid starvation become largely focused on acquiring and using a drug despite devastating consequences to health and well-being. These modern theories have distinguished between addiction and becoming physically habituated to a drug. Abusing caffeine, nicotine, or alcohol can lead to withdrawal symptoms once the person stops the drug, but those symptoms are distinct from the processes in the brain that cause addiction. Cigarettes are powerfully addictive, for instance, but smokers are spared the violent withdrawal symptoms that alcoholics and heroin addicts commonly experience.

Nineteenth-century physicians did not make this distinction between habituation and addiction. Instead, most saw the two as firmly linked. After 1813 the increasingly common experience of treating inebriates for what we would today identify as acute withdrawal symptoms made the drunkard's "overwhelming compulsion" newly and dramatically visible. As withdrawal symptoms began to appear, for instance, physicians commonly struggled with patients who begged for liquor to stave off delirium tremens. Writing in 1819, Philadelphia physician Gilbert Flagler related a case history of a patient who initially seemed to be suffering from a stomach ailment. On the second day of treatment, the patient "begged to have some gin, and said, that unless he could have it, he was apprehensive of a fit of convulsions." Flagler gave him a "small glass," but the patient continued to beg for more. Increasingly concerned, Flagler gave him gin and opium, but these doses failed to stave off the onset of delirium tremens. Over the next week, liquor's hold over the man was dramatized by his delusions and maniacal fits. He spent nights raving, "very furious," and threatened to kill his wife. These wild behaviors accompanied other symptoms such as vomiting up "great quantities of blood," violent diarrhea, and periods of "epileptic convulsions."[13]

Similar tugs-of-war between physicians and alcohol-craving patients became more common in the 1820s as the diagnosis of delirium tremens was more widely used. Before the 1820s, caregivers would have granted the patient alcohol from the beginning, without a second thought. As physicians responded to the new delirium tremens diagnosis with treatments focused on vomiting and opium, however, they became increasingly reluctant to give patients alcohol, both because they were aware that heavy drinking had created the condition in the first place and out of a conviction that all their patients should be discouraged from drinking liquor. This impulse to deny liquor to heavy-drinking patients likely increased incidents of acute withdrawal. It also made the drunkard's compulsion to drink more visible.

In daily medical practice, delirium tremens patients stood out not just for their incessant and desperate cravings, but also for their recidivism. The new delirium tremens diagnosis made these patients' relapses quantifiable in hospital records. From 1823 to 1850, for instance, the admittance records of the Pennsylvania Hospital recorded 750 people admitted with delirium tremens.[14] At least 265 of these admissions were patients who had previously been treated for the disease. One patient inflated this number significantly, accounting for 25 percent of these 265 cases. Between July 1, 1827, and July 10, 1850, Joseph Calhoun, a shoemaker, was treated for delirium tremens sixty-five times. Before the delirium tremens diagnosis, Calhoun would likely have been repeatedly jailed as a drunkard. Instead, he was diagnosed and counted as suffering from a physiological and psychological affliction.

Daily interaction with delirium tremens convinced physicians who worked in medical wards that drinkers could become physically habituated to alcohol and that it was the sudden abstraction of liquor that caused the violent symptoms. In his influential essay on delirium tremens, Benjamin H. Coates wrote that "this disease is the result, not of the *application*, but of the *sudden intermission*, of the use of these articles," and further, Coates argued that delirium tremens followed only long-term "habitual" drinking: "In every instance, it has either occurred from the sudden change of a fixed habit, or at the abrupt termination of a debauch . . . of long continuance."[15]

Coates's assertion that delirium tremens was caused by a "sudden intermission" remained controversial among the temperance societies, which had begun advocating total abstinence from liquor. In 1831 the Pennsylvania Society for Discouraging the Use of Ardent Spirits argued that "the general impression, that it is unsafe for confirmed drunkards, or even for persons who have long indulged in the temperate use of ardent spirits, to relinquish it at once, is erroneous."[16] Other doctors maintained that delirium tremens could result from varied circumstances, sometimes when habitual heavy drinkers suddenly abstained from alcohol or when individuals not accustomed to drinking engaged in a heavy binge.[17] Physician William Sweetser believed that when counseling longtime heavy drinkers, the threat of delirium tremens did mean that "it might not be prudent to enjoin a sudden and entire abstinence from the use of spirituous drinks," at least with the extreme cases. In his experience, "it is no easy task to persuade an intemperate old man that his health does not require ardent spirits." However, if a patient does quit, in most cases "health is often a good deal improved," provided the patient "is not very aged, or the constitution too much shattered."[18]

What stands out in Coates's description of alcohol addiction is his conviction that it was irrefutably observable. He focused attention on the brain, reasoning that on consuming liquor or opium, "a great depression" of the mental faculties "is the leading effect" and that "the intellectual faculties are deprived of their usual accuracy, and after a very short interval, somnolency and a general diminution of all the mental powers, are the unfailing successors." Thus, over years of habitual drinking, the patient becomes "accustomed to the impression of an agent which diminishes the activity of the mind; it learns to obtain an approach to the healthy equilibrium, by resisting this narcotic; and, upon its sudden removal, passes immediately into a condition of inordinate action." Physicians may differ on what they call this condition, he wrote, according to the various doctrines and principles they apply to it. Regardless, that habitual heavy drinkers become physically dependent on alcohol is a "fact, having its existence in nature" and not based on abstract theorizing. The editors of the *Medical Recorder* called Coates's description of the pathology of intemperance "not only ingenious, but, we think, entirely consonant with those laws which govern the animal economy."[19]

Many accepted Coates's general assertion that compulsive drinking was a physical disease but rejected his emphasis on the brain. Attention increasingly focused on the stomach or, less often, the nervous system as a potential center of the pathology. Philadelphia physicians John Bell and

William Darrach argued that habitual drinking impairs "natural thirst, and eventually natural hunger," which drives the inebriate to "drink again and again, and gradually more and more frequently, and stronger and stronger draughts." Eventually "the stomach itself has become diseased by the artificial stimulant."[20] The consequences of this diseased state were thought to go well beyond simply rendering drinkers powerless to control their impulses. By depraving the stomach tissue, alcohol threatened virtually all of the body's systems and functions.

The Fatal Bowl

By 1830 medical interest became so broad that Philadelphia's College of Physicians and Surgeons introduced a regular course on the pathology of intemperance.[21] Through speeches, pamphlets, and other publications linked with various temperance and health reform organizations, physicians translated this new medical knowledge about the physical effects of heavy drinking for a lay audience. A report written by members of the Philadelphia Medical Society and published by the Pennsylvania Society for Discouraging the Use of Ardent Spirits argued that physicians had a moral obligation to share their knowledge: "Physicians unquestionably possess greater opportunities for bestowing useful advice on this subject than most other citizens. It is frequently their solemn and imperative duty to forewarn the individual, who tempts the fatal bowl, of the danger he is incurring to his health and his existence."[22]

For physicians active in the temperance movement, the pathology of intemperance served as wonderful propaganda. Drawing from medical literature, physicians made a range of striking claims about the dangers of even casual drinking. Widely circulated through the extensive networks established by temperance organizations, physicians' literature was heavily influential in shaping new representations of pathological drinking in American popular culture.

Dire medical warnings about pathological drinking focused overwhelmingly on men, even though doctors regularly encountered female drinkers in hospitals and in private practice.[23] Municipal and institutional records from Philadelphia demonstrate that although most patients diagnosed with delirium tremens were men, women also suffered the disease. In addition, poor women were commonly recorded as victims of "intemperance." At the Pennsylvania Hospital in the 1830s, in a few rare cases physicians attempted to cure respectable women of the habit of drinking.[24] But as I noted in chapter 3, physicians were well aware of the social stigma attached to women's drinking and protected their patients' social standing, sometimes by recording an opaque diagnosis for an alcoholic disease. In the medical warnings contained in temperance literature, women are mentioned only occasionally, and then most often in relation to family and childbearing. The Philadelphia Medical Society lamented that "a portion of the still-born children receive their death from the intemperance of the mother."[25] Female drinking also appeared as part of the larger concerns among some physicians that prescribing liquor as a medicine introduced their patients to a deadly habit. Dr. Harvey Lindsly wrote, "More female drunkards are made by this means than by any other."[26] For the most part, however, female drinking was something physicians were aware of but were reluctant to discuss openly.

A wave of essays, addresses, dissertations, journal articles, and resolutions by physicians and medical societies warned men that even casual tippling put them at risk for developing an "imperious craving" for liquor.[27] Widely excerpted in magazines and newspapers, the Philadelphia Medical Society report argued that intemperance was a progressive biological condition in which the body becomes habituated to alcoholic stimulation:

> Ardent spirits are notorious for the facility with which the human frame becomes familiarized to them; and, in order to renew the sensation enjoyed at first, it becomes indispensable to increase the dose. This change takes place by such slow degrees that the patient is seldom aware of the fact, and finds himself subjected to an imperious craving, where he fancied he was only enjoying an indulgence capable of being regulated by a proper discretion.[28]

All drinkers risked developing morbid cravings caused by diseased internal organs, which had the potential to drag them into dependence and depravity.

Physicians described liquor as having the power to mesmerize drinkers and compel them to frightening acts of self-destruction. In an 1827 address to the New Hampshire Medical Society, Dr. Reuben D. Mussey referred repeatedly to the "magic" power of alcohol to cast down even the most promising and respectable into poverty and disgrace: "You have seen the man of talents, industry and extensive usefulness . . . thrown down, by the magic power of alcohol, from the pinnacle of his elevation, to become the object of popular derision and abuse." Like Coates, Mussey based his description of the "witchery" of alcohol on the new theories of physiological medicine that were transforming medical practice in the 1820s:

> What is the secret of this witchery which strong drink exerts over the whole man? . . . After being received into the stomach, it is sucked up by absorbent vessels, is carried into the blood, and circulates through the alimentary organs, through the lungs, muscles, and brain, and doubtless through every organ of the body. Not a blood vessel however minute, not a thread of nerve in the whole animal machine escapes its influence.[29]

Mussey included dramatic descriptions of the power of this physiological desire, including the account of a drunkard confined to the almshouse. Repeatedly foiled in his attempts to procure liquor, the man became so desperate that he went into the "wood yard of the establishment, and placed his hand upon a block, and with an axe in the other, struck it off at a single blow. With the stump raised and streaming he ran into the house crying 'get some rum, get some rum, my hand is off.'" The attendants unthinkingly brought a bowl of rum, which the man thrust his bleeding arm into before quickly drinking the liquor, announcing, "Now, I am satisfied."[30]

The power of stories like Mussey's was heightened by placing them alongside harrowing accounts of delirium tremens and the spontaneous combustion of drunkards. Physicians often presented these side by side, in graphic detail, to dramatize the horrors of intemperance. In a small volume of temperance addresses delivered in Cincinnati in 1828, for instance, Daniel Drake included five appendixes; the second was a twelve-page article on spontaneous combustion, and the third reprinted large portions of an article on delirium tremens that he had first published in a Philadelphia medical journal in 1818.[31] Publicizing delirium tremens and the spontaneous combustion of drunkards was part of physicians' larger campaign to share medical knowledge that could persuade those already temperate to remain so. In the marketplace of American popular culture, however, delirium tremens, the spontaneous combustion of drunkards, and other medical topics related to intemperance became subjects of popular speculation and proliferated in ways physicians had not intended.

Like delirium tremens, spontaneous combustion first appeared as an urban calamity during a period of growing anxiety regarding alcohol, poverty, and social order. European physicians had first documented the spontaneous combustion of drunkards in the seventeenth century, and interest grew in the first half of the eighteenth. These narratives almost always involved obese older women, often widows, who lived alone. Most often male doctors recorded the cases and shared them in learned societies and publications.[32] While many of the reported cases occurred on the Continent, the stories had special appeal in Britain during the "gin epidemic" of 1720–51, when the consumption of cheap spirits rose sharply.[33] Popular stories and illustrations from these years often portrayed solitary and impoverished widows who both drank and sold large quantities of cheap spirits. The gruesome stories of the charred bodies of old women evoked a cultural antipathy toward this particular class of women. In 1800 an article by Pierre-Aimé Lair in the French *Journal de Physique* revived interest in spontaneous combustion. Translated into English, the article appeared first in London and then, in 1812, in the Philadelphia journal *Emporium of the Arts and Sciences*, published by John Redman Coxe, a physician and a professor at the University of Pennsylvania.[34]

In the early nineteenth century, the remarkable claim was difficult for men of science to dismiss, even if most expressed doubts. Coxe published the article on the spontaneous combustion of drunkards, for instance, at a moment when nonhuman spontaneous combustion was a subject of much debate. In Coxe's *Emporium*, the article appeared among a succession of other essays documenting the spontaneous burning of substances such as wood, oil, and gunpowder.[35] Just the previous year, one of Philadelphia's most widely read daily newspapers, the *Aurora General Advertiser*, had also run a long editorial on the spontaneous combustion of construction materials.[36] During the 1820s, spontaneous combustion also appeared in several books by respected physicians, including the Britons Thomas Trotter and Robert Macnish.[37]

Some skeptics accepted the validity of the case histories but reasoned that the combustion might not be spontaneous. Heavy drinking might cause hydrogen gas, for instance, to accumulate in drunkards' fatty tissues. A spark or exposed flame then could ignite the gas, setting the body on fire. Macnish reprinted one typical case collected by Lair. Fifty-year-old Mary Clues had been "much addicted to intoxication" and became bedridden. She continued to drink and smoke despite her debility. One night, before going to bed, she placed two large pieces of coal on the fire and set a candle next to her bed. Early the next morning, "smoke was seen issuing through the window, and the door being speedily broken open, some flames which were in the room were extinguished. Between the bed and the chimney were found the remains of the unfortunate Clues; one leg and a thigh were still entire, but there remained nothing of the skin, the muscles and the viscera." Observers who entered the room were struck that while the body had been mostly consumed, there was relatively little damage to the room and its furniture. Macnish skeptically noted there was no firm evidence that this combustion was spontaneous. It could easily have been touched off by the candle or the fire.[38]

By the 1830s, spontaneous combustion appeared much more often in temperance literature, fiction, and newspapers than in formal science or medical publications. Aside from Coxe's *Emporium*, only one case of human combustion appeared in any of the leading American medical journals.[39] In temperance literature, physicians continued to cite spontaneous combustion as evidence of the harrowing dangers of drinking, but they prefaced these stories with the acknowledgment that "credulity seems tasked to believe their actual occurrence."[40] One popular account dwelled on the gory death of a Canadian man of twenty-five. The horrifying story played on the theatrical nature of the tragedy in comparing spontaneous combustion to eternal damnation. A known drunkard, the man was found one evening "literally roasted from the crown of his head to the sole of his foot." The doctor reported that he saw no "possibility of fire having been communicated to him from any external source. It was purely a case of spontaneous ignition. A general sloughing came on, and his flesh was consumed or removed in the dressing, leaving the bones and a few of the larger blood-vessels standing." A witness described the man as having "exactly the appearance of the wick of a burning candle in the midst of its own flame." The man survived thirteen days after the horrible event. The doctor reported, "His shrieks, his cries and lamentations were enough to rend a heart of adamant. He complained of no pain of body—his flesh was gone. He said he was suffering the torments of hell; That he was just upon its threshold and should soon enter its dismal cavern; and in this frame of mind gave up the ghost. O the death of a drunkard!"[41] The spontaneous combustion of drunkards infused alcohol abuse with a supernatural dimension.

Descriptions of delirium tremens were also among doctors' most emphatic assertions of the evils of drink. Unlike spontaneous combustion, in writing for the temperance cause physicians could rely on their personal experiences, a recent and extensive medical literature, and statistics showing the disease to be common in hospitals and almshouses. In his 1828 temperance address, Drake's account of the health consequences of alcohol abuse closed with a general description of the horrors of delirium tremens. Despite its religious and supernatural imagery, Drake's account mirrored those that had been circulating in medical journals since 1815. The drunkard's mind becomes disordered, Drake warned, and

> his imagination excited to an unnatural degree. . . . He occupies himself solely with the creations of his own distempered fancy . . . At one hour he converses with friends who are absent—at another . . . is enraptured with the

> sight of imaginary groups of celestial visitants, moving through the air; but in the midst of his bright enchantment, sudden darkness surrounds him, and Satan, with a train of frightful goblins, passes dimly but fearfully before his distempered sight . . . the frightful images cling to him like the poisoned shirt of Hercules.[42]

Before 1827, physicians largely had confined discussion of the disease to medical publications and institutions. Now widely described in temperance literature, delirium tremens passed into popular culture, dramatizing the horrible consequences of alcohol abuse and painting a horrifying picture of the violent psychology created by habitual drinking.[43]

The further history of Dr. Sewall's stomach plates demonstrates the broad resonance of physicians' temperance campaign as well as its ambiguous consequences. In 1843 one of the wealthiest and most influential national temperance activists, New Yorker Edward C. Delavan, gave Sewall's theories and gruesome illustrations a national audience. Delavan first published a small black-and-white composite image of the stomach drawings in the *Temperance Recorder*, a nationally circulating journal (see fig. 8). Separately, he published life-sized color images of the individual stomachs, with detailed annotations describing the tissue damage. The bound booklet also included an essay by Sewall summarizing his theories on the pathology of drunkenness. Delavan also had reproductions printed that were large enough to be displayed in museums.[44] A skillful propagandist, he disseminated the images through the many networks of the benevolent empire. He embarked on a campaign to send a copy of the booklet to every public school library in New York State. The superintendent of common schools wrote to Delavan, "I am satisfied that the colored plates of Dr. Sewall, exactly depicting the transitions of the human stomach from perfect health to the last stages of cancerous, alcoholic disease, will make a deeper and more lasting impression upon the minds of reflecting individuals and even upon the thoughtless and ignorant, than any other work that has ever been published."[45]

Additionally, The American Temperance Union claimed that General Winfield Scott was moving to post the plates in all American military installations, and that the presidents of the "Marine Insurance Companies expressed a wish that they might be put on board of every vessel on the ocean, on our rivers, and on our lakes, counteracting the peculiar temptations to which mariners and emigrants were exposed."[46]

The stomach illustrations also drew controversy, however, illustrating tensions that grew up between the medical community and moral re-

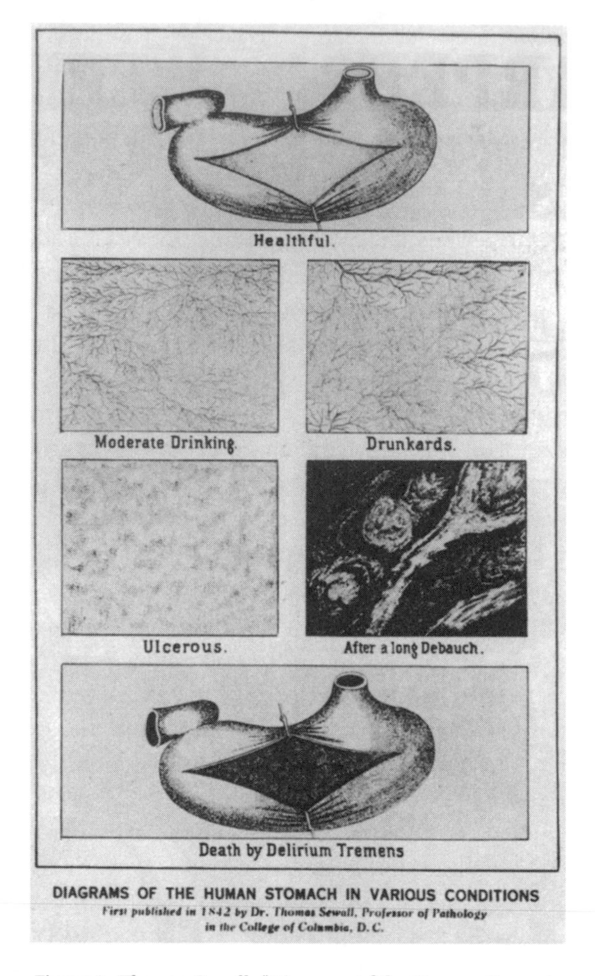

Figure 8. Thomas Sewall, "Diagrams of the Human Stomach in Various Conditions," c. 1842. From J. C. Furnass, *The Life and Times of Demon Rum* (New York: Putnam, 1965), 128.

formers who sought to use medical literature to further their own goals. In 1843, shortly after Delavan began trumpeting Sewall's stomachs, a long and wordy debate over the accuracy of the illustrations erupted in the pages of New York's *Evening Journal*. The controversy began when the faculty of the Albany Medical College voted not to display large reproductions of Sewall's stomachs in their public museum. As Dr. Thomas Hun explained, the faculty objected not to the idea that drunkenness created disease in the stomach, but to Sewall's portrayal of the various gradations of disease and the inaccuracy of some of the illustrations. Particularly, Dr. Hun objected to plate 2, "Moderate Drinking," which he said falsely advanced the view that "temperate drinking produces a disease differing only *in degree* from that produced by drunkenness."[47] Rather, he averred, "the habit of temperate drinking, which has prevailed at all times and in all places, which has been practiced by the best and wisest men, and the example of which has been given by the highest authority recognized by the Christian world, does not produce disease of the stomach." Nevertheless, he wrote, he had no great objection to most of the drawings and, perhaps in an attempt to mollify temperance activists, readily admitted that "the stomach is diseased in drunkards."[48]

A relentless advocate for total abstinence from all alcoholic drinks, including communion wine, Delavan refused to let Dr. Hun off the hook. Delavan no doubt saw an opportunity in the controversy to further publicize the images and the medical doctrines supporting total abstinence. The debate allowed him to expound at great length on every aspect of the drawings—from the close attention to shades of red that the printer had employed, to what he portrayed as the vast amount of medical evidence in his favor. Hun did not appreciate his role in Delavan's publicity bid. As the debate dragged on for more than a month, Hun declared that he was exasperated. "When I look back I cannot help feeling ashamed to have spent so much time about so ridiculous a matter." Complaining about being "dragged before the public" in an "age of charlatanism," Hun wrote,

> In former days, all who opposed the inquisition were considered as enemies of the Christian church; during the French revolution, all who spoke against the guillotine were considered as enemies of liberty. And to come from great things to small, we find at this day, Joe Smith accusing all who refuse to contribute to the building of his temple of being opposed to the cause of religion; and others are again crying out that all those who are opposed to the disgusting exhibition, in every tavern and every steamboat, of the human stomach in a state of loathsome disease, are enemies of the cause of temperance.[49]

Hun's frustration at Delavan's excesses aside, their debate highlighted the fact that in the 1840s physicians still varied widely in their opinions on pathological drinking and the impact of alcohol on human physiology. Hun believed drunkenness was most strongly felt in the nervous system, not the stomach. Physicians were quick to describe intemperate drinking as a disease, but medical opinion was not close to a consensus on its pathology, something Delavan was not interested in considering.

Sewall's stomachs initiated a long history of anatomical temperance propaganda. In the late twentieth century, for instance, the blackened lungs of inveterate smokers became a common image in public education campaigns especially meant to frighten schoolchildren. But just as the smokers' lungs operate primarily to horrify, graphic images of drunkards' stomachs did little to demystify the self-destructive compulsion to drink alcohol. The physiological principles Sewall derived his theories from based their truth claims on empirical evidence. Displaying the damage to the stomach lining was meant to plainly and dramatically illustrate alcohol's power to disease the body. But in translating the latest medical science for temperance publications, some members of the medical profession couched the problem in terms much more stark, disturbing, and pessimistic than anything portrayed by eighteenth-century authors, most notably Benjamin Rush. By declaring their allegiance to a new scientific openness and empirical evidence, physicians popularized a view of intemperance that celebrated the dark mystery of the drunkard's compulsion to drink.

Just as with spontaneous combustion and delirium tremens, Sewall's stomach images lent themselves both to moral instruction and to lurid sensationalism. Less than two years after their first publication, the stomach images had been made into a magic-lantern show. In Philadelphia, a Mr. Rodgers advertised his lecture in the temperance publication *Cold Water Magazine*. Rodgers had been conducting lectures "on the manners and customs of the Hindoos," but in Philadelphia, "in addition to his former paintings he now exhibits the representation of the drunkard's stomach, in four different stages of disease—from that of a perfectly healthy state to that of death from delirium tremens or mania a potu, where the stomach presents a most repulsive view of decay and mortification." The "repulsive" decayed and mortified stomach of a delirium tremens victim was a topic of extraordinary interest in some way compatible with an orientalist lecture on the "customs of the Hindoos."

Mr. Rodgers was not alone in recognizing the allure of Sewall's stomachs. No later than 1850, images of the drunkard's stomach were mass-produced as magic-lantern slides by the T. H. McAllister Company, one of the largest of such nineteenth-century American companies (fig. 7). The eerie power of the magic lantern revealed the gothic horror that underlay the stomach images in particular and the science of anatomy and physiology more generally. The illustration invites observers to peer like voyeurs into the darkest recesses of the drunkard's pathological nature. As Mr. Rodgers's advertisement read, "No drunkard can look upon these paintings, as represented by the Magic Lantern, without being horror struck at his own degraded situation."[50] The belief that contemplating the putrid organs of inebriates would have moral benefits for schoolchildren, soldiers, sailors, or middle-class men rested as much on the inherent horror of death and dissection as on the graphic depiction of the health effects of drink.

An Incurable Affliction

Hammering at the astonishing dangers of liquor consumption, the physicians most involved in articulating these claims also, at least initially, rejected any hope of a cure, even though new approaches showed some promise. Instead, they drew on medical science to construct liquor's power as a form of "magic" and "witchery." New disease models are generally advanced to bring greater understanding and thereby shape new treatment strategies. Why would leading physicians reject the possibility of a cure?

One answer is that despite broad acceptance of these new theories, most physicians still believed it was impossible to cure inveterate drinkers. Writing in 1827, Dr. Samuel Emlem of the Pennsylvania Hospital captured the general medical sentiment when he wrote, "The utter hopelessness of the reformation of the habitual inebriate" is a "striking and lamentable fact," recognized by "every observing member of the medical profession."[51] This medical opinion also fit with the stated goals of temperance organizations founded in the late 1820s. These organizations were not interested in reforming confirmed drunkards. Rather, they sought through "moral suasion" to persuade those already temperate to remain so. Even as the medical profession became heavily involved in treating delirium tremens, most doctors did not consider treating the more general problem of compulsive drinking as a medical problem.

But this sentiment ran counter to popular demand, which grew as the temperance movement and the widespread publication of the health consequences of heavy drinking created a new emphasis on sobriety as a marker of physical health and social respectability. More and more people discovered that kicking the habit was easier said than done. Even the discovery that drinking alcohol was itself a habit, rather than just part of a daily diet, was a revelation to many. In response, chemical curatives for chronic drinkers became commercially available just months after the flowering of temperance organizations in 1826. Despite some promising results, leading physicians and medical organizations were extremely skeptical.

Dr. William Chambers's Remedy for Intemperance, or "Chambers's Nostrum," as some physicians called it, was the most popular curative. A professor at Rutgers Medical College, Dr. Chambers marketed his remedy as a patent medicine and sold it directly to consumers through a network of pharmacists, grocers, and other dealers. His advertisements began to appear in early 1827, just months after the founding of the national American Temperance Society. Dr. Chambers died that same year, but James Hart, a New York physician, and A. M. Fanning, a dry goods dealer, quickly secured the rights to distribute the remedy and continued to sell it.[52] Many temperance organizations helped advertise Chambers's Nostrum. Advertisements appeared in publications from Boston and New York to Philadelphia, Tennessee, and as far west as Chicago.[53]

Dr. Chambers based his remedy on the belief that intemperance derived from a diseased stomach. The medicine contained an emetic that, according to reports, acted swiftly and sometimes powerfully when mixed into the drunkard's favorite liquor. Dealers selling the nostrum offered predictably optimistic testimonials, including one from a Pittsburgh physician who wrote of astounding success. He first tried the medicine on a confirmed drunkard so set in his habits that it took several days to sober him up enough to take the medicine. The next morning, the physician visited the man, who "met me with pleasure beaming in his countenance and said, 'I am cured.' On questioning him relative to this matter he assured me that early that morning on paying the usual visit to the tavern bar room, the smell of whiskey & the breath of those who were drinking, immediately met him with an unpleasant and disgusting effect, which he had never noticed before since he first commenced drinking."[54]

The close association of poverty and intemperance led at least one reformer to attempt the widespread use of Chambers's Remedy as a way of lifting individuals out of poverty. Boston's first Unitarian minister to the poor, Joseph Tuckerman, came to believe that the principal cause of destitution was the degrading habit of intemperance and that drinking greatly increased the suffering of the poor.[55] Convinced that the most effective way to combat poverty was to free men from intemperance, he bought large quantities of the medicine and enthusiastically administered it to men in his ministry. Despite some exciting early successes, however, the medicine made several of Tuckerman's charges violently ill. He abandoned his efforts by the end of the summer.[56]

Physicians largely dismissed Chambers's popular remedy. Their antipathy derived in part from reports that the medicine could be dangerous. The Medical Society of the City of New-York declared that Chambers's Remedy had "ruinous tendencies" and "frequently fatal effects." But physicians' opposition to the cure was not based entirely on medical concerns. They admitted, for instance, that the medicine did seem to work in many cases, and that the health consequences of drinking were so dire that the risks might be worthwhile.[57] Further, Chambers's Remedy was not a great departure from other contemporary medical therapies. By acting to cleanse the stomach, it was certainly consistent with current models of the pathology of alcohol addiction. Physicians also commonly used strong emetics on patients. The analysis of the New York Medical Society found the main active ingredient to be tartar emetic, the same ingredient Dr. Joseph Klapp had championed as a treatment for delirium tremens in 1817. While Klapp's cure had also been attacked as dangerous and potentially deadly, tartar emetic was firmly in American physicians' pharmacopoeia, and some had experimented with it as a cure for drunkenness. In one of his lectures to medical students at the University of Pennsylvania, none other than Benjamin Rush reported that he had used the compound to cure one of his black servants of intemperance.[58] In 1827 a physician in rural Pennsylvania reported privately to Benjamin H. Coates that he had administered tartar emetic to an intemperate patient without his knowledge.

> A drunken rogue, who has given me a vast deal of trouble with his mania á potu, came to my shop half drunk: I knew what was to come of it the next day & I therefore gave him a cup of tea in which I olipt ten gras. Tart. Emet. [*sic*] He went a fishing with another person as drunk as himself & he lay on the bank puking & purging the whole night. This cured him of drink for three months & saved me much unpleasant attendance.[59]

Given their strong rhetoric regarding the horrific health consequences of intemperance, recommending a strong drug, even a potentially risky course of treatment, would seem justified.

What seems to have been far more pressing to the members of professional medical societies was that Chambers's Remedy was a "quack" medicine, marketed as a secret recipe and sold directly to consumers. The report on Chambers's Nostrum done by New York's medical society was part of a larger article demanding that the state legislature do more to regulate the practice of medicine and protect orthodox physicians from competition. "It is in the laudable pursuit of the great object of the suppression of quackery," the article read, "that the investigations . . . were undertaken." Their goal was to "enlighten the public mind . . . to unravel the mystery and dissolve the charm invested in secret remedies." And further, "exposing their ingredients and their powers of doing harm, is . . . an imperative duty."[60] Just as professional incentives encouraged physicians to be leaders in the temperance movement, these imperatives led them to oppose a popular treatment.

Whether because of opposition from the medical profession or growing alarm over the potentially fatal side effects, the popularity of Chambers's Remedy was relatively short-lived, and advertisements for the medicine disappeared by 1830. One factor may have been its price. A newspaper correspondent, responding to the findings of the New York medical society that the main ingredient was tartar emetic, wrote, "If the above report be correct, it follows that any of our Medical men can furnish a remedy equally efficacious as this celebrated remedy, for five cents, instead of five dollars, the price which is paid for this."[61]

The orthodox medical profession gave at least some scientific respect to a different chemical cure. A German doctor first advanced sulfuric acid as a cure for intemperance in 1818.[62] Like tartar emetic, the cure worked by inducing nausea. First tried in the United States by the Philadelphia physician William Brincklé, sulfuric acid was never promoted and sold commercially in the manner of Chambers's Nostrum. Brincklé published the results of his experiments in an 1827 article in Coates's medical journal. Despite some short-term successes, Brincklé acknowledged that the long-term results were mixed. He nevertheless insisted on the promise of the cure and called for establishing an inebriate asylum where drunkards could be treated systematically. By 1829 Brincklé's cure had received some attention, although no efforts were made to institute it on a wide scale. Sweetser noted Brincklé's experiments and wrote skeptically, "If this is true, it is very difficult to determine the mode in which [sulfuric acid] operates to produce such an important result."[63] Two years later, the Philadelphia Medical Society said in a report that while they did not doubt Brincklé's veracity, their own results had been "doubtful and disappointing."[64] Perhaps as a concession to Brincklé, who was a member of the society, the report did say that the sulfuric acid cure was worthy of further experimentation. Brincklé was allowed to include his own testimonial, and he reported that he had continued to administer the cure and had some success.[65] The sulfuric acid cure likely also had commercial potential, although Brincklé seems to have lacked Chambers's entrepreneurial instincts. His cure received very favorable reviews in popular journals, including the *New England Farmer*, which wrote in 1828, "It has recently been discovered, that sulphuric acid, taken in spirits, completely eradicates the inclination to use them intemperately. It is said to be preferable to Chambers' remedy, being more simple, cheap, and wholly innocent."[66]

The significance of Chambers's Nostrum and the sulfuric acid cure lies not so much in their efficacy as in their cool reception from elite physicians. After all, both treatments rested on current medical models of the pathology of intemperance, which argued that it was a disease of the stomach. In fact, both treatments resembled twentieth-century aversion therapy treatments, such as the drug disulfiram, also known as Antabuse. It also works by making users physically ill when they drink alcohol. First marketed in 1948, disulfiram remains available as a treatment for alcoholism, although it is not widely used.[67] While sulfuric acid and tartar emetic may have offered mixed results, at least one respected Philadelphia physician remained optimistic about the efficacy of sulfuric acid after more than two years of experimentation.

Physicians in the 1820s and 1830s chose to avoid treating intemperance not because they lacked potential therapies, but in large part because professional incentives were strongly weighted against it. Involvement in temperance societies raised physicians' public profile and identified them with a popular social reform movement. Giving speeches, publishing commentaries in newspapers, and disseminating medical knowledge held great benefits for individual physicians who aimed to establish and maintain a private practice. And physicians supported the message of early temperance societies, which generally eschewed trying to recover confirmed drunkards, focusing instead on persuading the temperate, especially young men, to remain so. It is also true that treating an inebriate in a private practice, outside an institution, was uncertain.[68] Given the universal availability and affordability of hard liquor, any patient subjected to sulfuric acid or a similar cure would have ample opportunity to relapse. This fact, along with the realities of the medical marketplace, meant that very few physicians were inclined to try to treat intemperance, even if most agreed it was a disease.

Toward Asylums

By midcentury, incentives were slowly changing with the development of stronger professional organizations, larger medical institutions, and especially the asylum movement. This movement would form the foundation for the rapid expansion of inebriate care after the Civil War. The impulse to construct institutions dedicated to the care of the mentally ill began to take form in the early nineteenth century. In Europe, especially Britain, France, and Germany, a growing number of physicians dedicated their professional lives to working in asylums. In the United States, however, asylum medicine was still decades away from becoming an important specialty.

Nevertheless, by the 1820s several asylums had been built, such as those at Worcester, Massachusetts, and Hartford, Connecticut.[69] The Pennsylvania Hospital also reserved an entire wing for mentally ill patients. In the 1820s, confinement in a hospital or almshouse medical ward was for the most part a hardship endured by the poor out of extreme necessity. Anyone with means preferred to pay a doctor to treat them at home. The new asylums were not intended for the poor alone, however. Promoted as a humane and effective treatment for insanity, moral therapy sought to shelter patients in a regimented but dignified environment that replicated the bourgeois family home. Eschewing physical coercion as much as possible, physicians sought to win patients' trust and consent for treatment, thereby strengthening their capacity for self-control. Daily routines included music, lectures, educational magic-lantern exhibitions, exercise classes, walks in ornamental gardens, and other diversions.[70]

The founding of the American Temperance Union in 1826 inspired advocates for mental asylums to argue that the new institutions might also be used to treat inebriates. In 1833 Dr. Samuel Woodward, superintendent of the mental hospital at Worcester, Massachusetts, argued for inebriate asylums in a series of essays in a local newspaper. Woodward shared the medical consensus that casual drinking could become an uncontrollable compulsion. "Like insanity," Woodward wrote, "intemperance is too much of a physical disease to be cured by moral means only. The appetite is wholly physical, depending on a condition of the stomach and nervous system, which transcends all ordinary motives of abstinence." Cravings create "the desire of immediate relief so entirely incontrollable [*sic*], that it is quite questionable whether the moral power of many of its victims is sufficient to withstand its imperative demands." Woodward recommended that "confinement and restraint" were "absolutely necessary for a cure, till remedies can be applied to remove the physical suffering, and bring the subject of it within the range of moral influence."[71] This view of the affliction as both moral and medical was fundamental to the developing rationale for treating inebriates in asylums.

Although in popular discourse "intemperance" was often associated with the worst manifestations of urban poverty, depravity, and criminality, advocates for building new inebriate asylums saw potential patients coming from a respectable and affluent social milieu. Published as a pamphlet in 1838, Woodward's essays inspired a brief campaign in Philadelphia to build an inebriate asylum. Led by Philadelphia physicians and temperance advocates, including longtime physician-activist John Bell, the campaign began with a large public meeting in November 1840. They proposed that the city build "a respectable Asylum" for inebriates that would function as "a hiding place from their enemy—call it, if you please *the Retreat.*" In curing the intemperate, the goal was to preserve the idealized middle-class family. "How many of these persons are in the bosom of our families, surrounded by all its endearments and comforts, and yet who destroy the peace of those once happy homes." The class considerations of asylum advocates were readily apparent in the sentimental portraits they drew of potential patients:

> Look at yet another picture. There is a wife! Her house and children once showed care, economy and comfort; and her husband went forth to his occupation daily with hope and returned with pleasure. But now that home is neglected and disordered. For a while there was only a faint and silent suspicion. But, her frequent indispositions, her changed countenance and deportment, and the concealed vials about her dwelling, made it no longer a secret. Her children and husband may still be near her; but her affection for them is not so strong as for that which will deaden a craving appetite and the gnawings of a diseased stomach. Is there no remedy?

The proposed inebriate retreat was intended to cure middle-class drinkers and return them to the domestic embrace of their loved ones.[72]

While dwelling on this sentimental appeal, these advocates proposed to treat the affliction as a progressive physiological disease centered in the stomach, as physicians had been describing it since the 1820s. "*This is* Intemperance," they wrote:

> The constant and daily impression of so powerful an agent as alcohol, must produce a marked action, both of the nerves and vessels of the stomach. . . . The brain, the nerves, the heart, and vessels, the glands, all receive an impulse from its influence, and all get into a condition which renders this continued influence necessary. After a time, the quantity must be increased or the effect is lost, and more in quantity or stronger doses must be substituted.[73]

To counteract this compulsion, asylum advocates proposed combining a strict diet, founded on the principles of physiological medicine, with the "moral" therapy championed by the asylum movement. Promoters believed that an inebriate asylum organized around these routines could provide the drunkard with a strict routine and a carefully controlled diet of water and bland foods, which would rehabilitate the tissues of the stomach, thereby curing the cravings for alcohol while also instilling proper habits.

The brief Philadelphia movement failed, but physicians did experiment with treating pathological drinking first in the mental ward at the Pennsylvania Hospital and, more extensively after 1841, at the new Pennsylvania Hospital for the Insane, overseen by Dr. Thomas Kirkbride.[74] Intemperance was an important concern at Kirkbride's asylum, but more as a cause of insanity than as a subject of treatment in itself.[75] In the mid-nineteenth century, leading physicians believed that insanity could be caused by various forms of mental stress, of which intemperate drinking was just one. In his published annual reports, a table "showing the supposed causes of insanity" demonstrates that cases of insanity caused by "intemperance" rose faster than the overall population of the asylum, from 16 out of 176 total patients (9 percent) in 1841 to 170 out of 1,064 total patients (16 percent) in 1854.[76] In the 1847 report, Kirkbride listed thirty-five categories in this table, and "Intemperance" was second only to "Ill Health," which accounted for 190 cases. Other significant categories included 61 patients who had suffered "Loss of Property, failures"; 56 had been struck with "grief"; and 57 were driven insane by "mental anxiety." His 1847 report also listed cases caused by, for instance, mortified pride (3) disappointed expectations (8), and want of employment (19). Even "exposure to direct rays of the sun" and intense heat were seen as possible causes. Intemperance was far more common than related categories like opium use, which accounted for just one male and four female cases. Cases of insanity caused by tobacco (3) and masturbation (10) were found exclusively in male patients.[77]

Kirkbride believed that treating intemperance at the asylum was appropriate and necessary, but only for certain types of patients: those who were driven to insanity by heavy drinking or who exhibited an "uncontrollable fondness" for alcohol as a symptom of insanity. Other "quite numerous classes of habitual drinkers" were not appropriate for treatment, especially unrepentant drinkers susceptible to violence and criminality. Not only were treatments ineffectual with such patients, but "the moral effect produced on other patients in the wards . . . is almost always unhappy." But Kirkbride did believe that more respectable and remorseful inebriates had a place in his asylum. "When not under the influence of the habit," he wrote, these patients "are fully sensible of its enormity, and . . . are anxious to reform." This class of inebriates came from every walk of life:

> No business or profession is exempt, not even ministers of the Gospel. From the histories given by patients or their friends, it is common to learn that the sufferer is a man of liberal education, ample wealth, surrounded by an affectionate and devoted family, happy in all his domestic relations and respected

> in the community; himself a truly benevolent man, active in works of charity, and ever ready to assist the suffering, yet, with all this, an uncontrollable fondness for stimulation is destroying everything.

He only "occasionally" accepted such cases, however. Kirkbride was reluctant because he had found that owing to the strict discipline of the institution and the necessity of living and interacting with the patient population, these inebriates were very "likely to leave before their reformation was complete." Kirkbride believed an institution devoted exclusively to inebriate care would be more appropriate and successful.[78]

Despite his reluctance, Kirkbride was often pressed to accept these patients.[79] He wrote that habitual intemperance was "constantly brought to the notice of those who have charge of Hospitals for the Insane, by . . . the earnest appeals for advice in reference to this unfortunate class of persons."[80] Concerns about respectability and the reputations of drinkers and their families drove many of these appeals. In 1846, J. C. Hall wrote to Kirkbride that a "Gentleman" of his acquaintance had become "so enslaved" to alcohol that "destruction to himself and family must be the result." Hall continued, "He has himself suggested that he should go to some Hospital, where his diet and habits should be subjected to medical control." But Hall's appeal was delicate, fully aware of the social implications of his friend's becoming stigmatized. "He is not a drunkard," Hall assured; "he is an educated well behaved man and would give no trouble." Hall concluded his letter with a broad appeal,

> I have often met with men who have expressed a desire to be placed beyond the reach of temptation and have earnestly sought for some treatment that would cure the morbid desire. . . . I am convinced that compulsory privation of drink—Therapeutic remedies—moral enlightenment and encouragement with mental recreation and physical employment would return many a valuable member to happiness and health."[81]

By midcentury, friends and families increasingly called on the medical establishment to help them reform loved ones whose drinking habits created conflict and violated social norms.

Whereas several decades earlier physicians had associated the new disease delirium tremens with businessmen and "the sons of genius," Kirkbride and the inebriate asylum advocates were careful to distance their new institutions from that disease. Delirium tremens remained a daily occurrence in the city's medical wards, but Kirkbride was emphatic in his annual reports: "Cases of Mania a Potu are received at the Hospital *in the city only.*" No doctors now asserted that delirium tremens typically struck the "sons of genius." Reflecting on the frustrating efforts to reform inebriates at the Pennsylvania Hospital, however, Coates offered sympathetic descriptions that sounded very similar to earlier characterizations of delirium tremens patients. He asserted that typical patients included "men of superior education and ability" with the "feelings of a gentleman," and "very refined" women.[82] Coates's sympathetic descriptions further illustrate that the impetus for treating patients for compulsive drinking derived from motivations very similar to the class concerns that had driven physicians' original interest in delirium tremens earlier in the century.

Even though physicians carefully selected patients who seemed appropriate and amenable to treatment, intemperate patients were troublesome indeed. At the Pennsylvania Hospital, Coates explained that the only successful course of treatment involved locking inebriates in their rooms to prevent them from drinking.[83] In Kirkbride's asylum, it was not uncommon for patients to escape to get drunk.[84] Further, patients being treated for compulsive drinking appeared, when sober, to be in good mental health, making it difficult to justify keeping them incarcerated.[85] Some intemperate patients obtained a writ of habeas corpus to compel their release. While only a few took this course, it put a tremendous strain on Kirkbride and generated unwanted publicity. Providing legal counsel in one such case, a lawyer advised Kirkbride, "I do not find any enactment which would seem to authorize the confinement or detention of an habitual drunkard against his will for curative treatment or otherwise unless he be also of unsound mind."[86]

Asylum treatment for inebriates remained difficult, ineffective, problematic, and thus relatively rare despite strong public demand. Nevertheless, Kirkbride's efforts foreshadowed the rapid development of inebriate asylums in the postwar period. As historian Sarah Tracy has documented, it was after the war that physicians set out create a new medical specialty focused on curing intemperance.[87] Through midcentury, that project was neither possible nor desirable. In the early republic, most physicians aspired to genteel private practices catering to the respectable, not a career treating drunkards in a large medical institution. After the Civil War, changing professional imperatives and marketplace conditions would reshape inebriate care, but public demand for that care had taken shape decades earlier.

While physicians played a central role in arousing popular concern about pathological drinking, their inability or unwillingness to respond to that concern created a vacuum that was filled in part by the Washingtonians.

Fundamental to the Washingtonian mission was the belief that drunkards could be reformed. As the young lawyer Abraham Lincoln said in an address to a temperance meeting in 1842, "By the Washingtonians this system of consigning the habitual drunkard to hopeless ruin is repudiated."[88] The Washingtonian "conversational meetings" sought to use the confessed experiences of former drunkards to reform the habits of inveterate drinkers.[89] Decidedly plebeian in their affiliations, the early Washingtonians rejected the social and spiritual elites that had led the most important early temperance societies. Writing in 1846, a New York rum dealer named Benjamin Estes described the lecturers at an outdoor Washingtonian meeting in that city. He was amazed that "the drunkard could be saved if these men told the truth," since his own physician father had taught him that "the confirmed drunkard could not be saved."[90] Although the Washingtonians' movement faltered after 1845, their efforts helped galvanize public support for the development of asylums and inebriate homes.[91]

The Washingtonian "conversation meeting" also had a profound effect on popular culture, especially representations of pathological drinking. Alcoholic diseases and uncontrolled cravings were central to the drama of such narratives. Significantly, the Washingtonians told these stories in the first person. Previously, in medical case histories, temperance lectures, sermons, and advice manuals, sober narrators described tortured cravings or alcoholic hallucinations suffered by others. Crucial to the appeal of Washingtonian meetings and the cultural productions that grew out of the movement were the compelling first-person narratives offered by former drinkers. Sincerity, emotion, and authenticity were the hallmarks of a powerful lecture, and graphic descriptions of the physical and psychological suffering of drunkards were a mainstay of the genre. Delirium tremens was a favorite topic.[92] Washingtonian narrators drew on the disease to dramatize the frightening depths of depravity they had reached. No lecturer was more famous than John B. Gough, whose public performances featured a dramatic reenactment of his own experience of delirium tremens. His training as a stage actor and singer no doubt helped him render the disease more poignantly. The frontispiece to his autobiographical *Platform Echoes*, repeatedly reprinted in the nineteenth century, includes a depiction of the author struggling with his mind's demons (fig. 9).

The sensational tales shared in conversational meetings quickly found their way into print. Novels written by lecturers or inspired by the Washingtonian societies took the form of confessional autobiography, relating a series of life events and crises caused by demon rum. Job loss, the death of a spouse or child, violence, criminality, depravity, and health crises were

Figure 9. John Gough, *Platform Echoes, or Leaves from My Notebook of Forty Years* (Hartford, CT: A. D. Worthington, 1886).

all common subjects. Authors commonly devoted chapters or more to their experience with delirium tremens, in such novels as *Autobiography of a Reformed Drunkard* (1845), *Confessions of a Reformed Inebriate* (1848), *The Glass, or The Trials of Helen More, a Thrilling Temperance Tale* (1849), *Six Nights with the Washingtonians* (1842), Gough's *Autobiography* (1845), and *The Horrors of Delirium Tremens* (1844), among others.

In these narratives, physicians often appear at moments of extreme crisis. In T. S. Arthur's *Six Nights with the Washingtonians*, one of the earliest and most successful temperance novels, a young man describes the serpents, monsters, devils, and other "strange terrors" that tortured him. Finally collapsing into sleep, he awoke to find himself under the care of a physician:

"What *has* been the matter with me, doctor?" I asked, after I was able to go about.

"*Mania-â-Potu*," he replied, in a low emphatic tone—

"Mania, what?" I said for I did not understand him.

"Mania-â-Potu," he repeated."

The physician explained that "when, by a long continued resort to artificial stimulus, anyone has weakened, to a certain degree, the vital energies of his system, the stimulus itself at last fails to keep up the apparently healthy action, and all things fall into disorder."[93] The physician educates the young man while delivering his diagnosis as a judge issues a sentence: drunkard.

Narratives often referred to the drunkard's pathological thirst for liquor. In Gough's *Autobiography*, which no doubt offers a taste of his famous public performances, he blends imagery from medical temperance literature—delirium tremens, fire, and overwhelming cravings—to create a frightening and dramatic scene. After a prolonged period of drinking, he felt "an awful sense of something dreadful coming upon me." Augmenting his sense of foreboding, he began hearing strange disembodied voices. At the same time, "The horrible, burning thirst was insupportable, and, to quench it . . . I clutched again and again, the rum-bottle, hugged my enemy, and poured the infernal fluid down my parched throat. But it was of no use—none." In desperation he began to smoke tobacco, hoping the "narcotic leaf" would calm him. Slipping briefly into sleep, "I awoke, and discovered my pillow to be on fire!" His neighbors rescued him from his burning mattress and certain death, but his affliction only continued. Growing desperate,

> I begged the people of the house to send for a physician . . . but I immediately repented having summoned him. . . . He saw at a glance what was the matter with me, ordered the persons to watch me carefully, and on no account to let me have any spirituous liquors . . . then came on the drunkard's remorseless Torturer—delirium tremens, in all its terrors, attacked me.[94]

Gough goes on to detail the vivid hallucinations characteristic of the disease. For him and other writers and lecturers, medical temperance literature offered a narrative structure and theatrical material to elicit strong emotions of shock and horror.

Here and elsewhere, the doctor appeared as a figure of calm authority, a representative of temperate rationality in a moment of overwhelming irrationality. In *The Glass*, a visit to the lunatic asylum foreshadows the female protagonist's own case of delirium tremens. On entering the cell of a man dying of delirium tremens, she describes the horrific scene.

> There are the keepers, grim and impatient, with countenances showing vexation at the trouble imposed on them, and lips kept close together, as though to restrain the maledictions which strove to escape. There stands the young physician, with a cold, calm countenance, witnessing without emotion an accustomed scene. There lay the wretched drunkard occasionally passive and silent, with the white foam escaping from his lips, his eyes glassed and staring, the pupils extended to their utmost dilatation, the blood seeming ready

> to start from his skin, his fingers quivering nervously—then, with fearful howls and shrieking curses, struggling in the strong arms of his captors.

A physician also attends her when she later succumbs to the disease. He stands by as an impassive presence when she discovers that her son has died because of her neglect while she was in the throes of insanity. Amid this alcoholic horror, authors portrayed physicians as knowing witnesses, authority figures hardened to the pathetic scenes.

Physicians' campaign to raise the profile of their profession by spreading knowledge about the pathology of intemperance thus had deeply ambiguous consequences. These novels generally portray physicians positively, but the popular authors also plumbed the lurid depths offered by medical descriptions of alcoholic disease, especially delirium tremens. As was typical of the medical literature, popular authors dwelled on descriptions of vivid hallucinations. These first-person narratives had a mesmerizing immediacy beyond third-person descriptions. One inebriate confessed that in his disease,

> huge serpents, with fiery eyes, and darting forth forked tongues, coiled upon the posts of my bed, and seemed ready to pierce me with their sharp white fangs. Then, perhaps, wherever I turned my eyes, horridly distorted faces would be looking at me, and perpetually assuming new, and disgusting and repulsive forms. Sometimes, ghastly skeleton shapes would peer in upon me from behind the half-opened curtains. Again, the scene would change, and white sheeted specters would glide in, and hover around me.[95]

Once loose in popular culture, delirium tremens became a topic for prurient imagination and sensational speculation.

Physicians lost control of the representations of pathological drinking and alcoholic diseases that had been central to their temperance campaign. Perhaps the most extravagant example of this loss is the evangelical James Root's *Horrors of Delirium Tremens* (1844). The book combines a vivid description of his experience with an extensive argument that delirium tremens was in fact not a disease, as physicians claimed. In his view, delirium tremens constituted a visitation by Satan and his minions.[96] He devoted most of the 483-page book to his argument that the devils that tormented him were in fact real, and he attacked physicians and others who claimed delirium tremens was simply a "derangement of his nervous systems." Mocking secular rationalists, he wrote, "It might be a great benefit to these little sceptics, if their nerves would only create a few devils. . . . For one attack from the devils would certainly drive their puerile and self-destroying skepticism out of their heads."[97] For Root, the real cure for the affliction was Christian salvation.

Physicians' efforts to warn of the dangers of drink thus went well beyond a campaign of moral suasion and advocating self-control. They dwelled on the sensational: alcoholic somnambulism, harrowing insanity, preternatural fires, and putrid organs. These images encouraged the public view that while temperance was crucial to health and well-being, drinking could easily develop into a powerful affliction. Popular culture amplified these sensational stories. In explaining the pathology of intemperance, the medical profession thus did not settle on one central diagnosis or even a single syndrome. In the mid-twentieth century physicians and public health officials would focus their efforts on the "alcoholism" diagnosis; in the antebellum era, the concern that heavy drinking could become an "imperious craving" was only one frightening consequence of heavy drinking. After midcentury it increasingly became the dominant concern. The drama of the drunkard struggling with his depraved impulses became a growing public and medical preoccupation as temperance carried ever greater social significance. The profession was not prepared to answer that challenge, however. The irony of physicians' campaign of moral suasion was that by publicly asserting medical authority and offering empirical evidence of the pathology of intemperance, they only fueled fascination with the power of demon rum to mesmerize, deprave, and ultimately destroy.

SIX

The Drunkard's Demons

In Joseph Allison's poem *The Rum Maniac,* a man slipping into the clutches of delirium tremens begs a physician to give him rum, becoming increasingly frantic as he is beset by horrors.

> But, Doctor, don't you see him there?
> In that dark corner low he sits:
> See! How he sports his fiery tongue,
> And at me fiery brimstone spits!

But even devils and the threat of eternal damnation cannot cure his compulsion for drink.

> Say, don't you see this demon fierce?
> Does no one hear? Will no one come?
> Oh save me—save me—I will give
> But rum! I must have—will have rum.[1]

Dragged down by his personal demons, the man sinks into the world of the damned. His lone consolation is that in hell, rum is never in short supply.

Published in temperance periodicals and pamphlets, *The Rum Maniac* is one illustration of how delirium tremens shaped representations of pathological drinking in the popular culture of mid-nineteenth-century America. It takes as its premise the compelling question that lay at the heart of the cultural fascination with the inebriate: Why does the drunkard continue to drink despite knowing all the dangers and horrific consequences?

Figure 10. Joseph Allison, *The Rum Maniac* (1851).
The Library Company of Philadelphia.

The supernatural beings express the dark mystery of the rum maniac's irrational compulsion, which the author paints as a Stygian suicide.

At midcentury, delirium tremens had become a topic of popular entertainment. The disease that had first captivated the imagination of Philadelphia medical students and their professors in the 1810s emerged from medical journals and temperance literature to haunt American theater and print. Representations of the disease can be found in the writings of some of the most enduring authors of the nineteenth century, such as Edgar Allan Poe, Walt Whitman, and Mark Twain, and in the era's most popular fiction by authors such as T. S. Arthur, Augustine Duganne, and George Lippard, whose *Quaker City, or The Monks of Monk Hall* (1844) was the best-selling novel before the publication of *Uncle Tom's Cabin* (1852). On stage, two of the most popular melodramas in all of nineteenth-century American theater, *The Drunkard, or The Fallen Saved!* and *Ten Nights in a Barroom*, cemented the horrors of delirium tremens in popular consciousness. At the climax of both dramas, white male actors graphically enacted the affliction's distinctive terrors. In these cultural productions, delirium tremens appeared less as a disease than as a compelling spectacle that illuminated the psychic power of intoxication and the compulsive nature of heavy drinking.

During decades when temperate habits had become firmly entrenched in the bourgeois requirements for respectable comportment, the broad visibility of the disease in middle-class entertainment is striking. Temperance societies were more numerous than ever before, and anti-alcohol forces were increasingly potent politically, exemplified in 1851 when Maine became the first state in the Union to ban the sale of alcoholic beverages. Why then were middle-class men and women flocking to theaters to watch graphic enactments of violent alcoholic insanity? What made literary and theatrical representations of these horrors so alluring? To say that novelists, poets, playwrights, illustrators, and editors drew on delirium tremens for sensational effect only skirts the question: What was the nature of this sensationalism? What meanings did audiences derive from these performances?

In 1855, one case history published in the *American Journal of the Medical Sciences* stands alone for its gruesome imagery, but it nevertheless makes explicit some of the symbolic meanings of the disease at midcentury. Dr. William T. Taylor described the case of a Philadelphia cigar maker who developed delirium tremens after an extended debauch. Late one night, "imagining that his relatives were accomplices of a crowd of demons," he fled from his home and "ran towards Girard College, intending to hide in a small wood near by." The college had been named for Philadelphia's patron saint of the American self-made man, the fabulously wealthy Stephen Girard. In his will, Girard had intended it to educate orphaned and impoverished boys and put them on the road to social betterment and economic success. At this highly symbolic site, the frantic cigar maker found only more horror. In the wood "he was met by a greater number of fiends, who, having caught and secured him, told him that, to appease their anger and obtain his liberty, he must sacrifice his virility." Finding a piece of a broken porter bottle, the man succeeded in cutting off his penis and testicles, working "three-quarters of an hour in excising the parts." Unmanned by liquor, the drunkard cigar maker was found sitting in a pool of blood chewing on his "lacerated and bloody organs." Refused admission to the more respectable Pennsylvania Hospital because his wound was the result of mania a potu, the man survived his injuries but died of pneumonia at the almshouse, highlighting the utterly squalid nature of the drunkard's death.[2]

More gothic horror story than case history, this narrative illustrates, in part, that physicians writing for publication, even if only for an audience of fellow physicians, were not writing in an insulated medical sphere, solely for the advancement of science. Case histories could also serve as prurient entertainment and moral commentary, even in leading medical journals.[3] Filled with compelling symbolism, the story demonstrates that delirium tremens held a dark allure for this author. Offering little information relevant to treating the affliction, his central concern is masculine failure.[4] Fleeing from his family, the cigar maker's self-castration graphically epitomized the destruction liquor had wrought on his patriarchal authority. The essay marvels at the power of the hallucinations that drive him to a shocking act of self-debasement. So mesmerized, so completely in the clutches of his own alcoholic demons, he endures self-mutilation and horrific pain.

Even the physicians who had initially defined it thus approached delirium tremens not just as a pathology, but also as a type of theater, dwelling as they did on the terrifying phantasmagoria that haunted their patients. As masculine success became increasingly defined as a lifelong moral project of self-cultivation in which the individual adhered to a strict regimen of healthy habits, hard work, Christian piety, and moral self-restraint, such stories of unsuspecting drinkers becoming slaves to grim apparitions and depraved alcoholic impulses proliferated in popular culture. Delirium tremens shaped a new theatricality of pathological drinking that emerged in the dark shadows cast by Girard, Franklin, and the rest of the era's archetypal self-made men. The disease became a performance in which the strict imperatives of a new market economy and its requirement of a rational, self-directed economic actor were cast off in a descent into an irrational nightmare realm of hideous demons.

This perverse fascination with delirium tremens highlighted deep conflicts in American attitudes toward drinking and, more broadly, the requirements for social advancement. Popular narratives commanded both horror and fascination, casting inebriates as romantic figures struggling with the power of their dark and diseased imaginations. Literary, poetic, theatrical, and visual representations of the disease most often centered on white men's struggle to attain respectability and defend their middle-class status. Narratives that detailed struggles with supernatural beings epitomized the drunkard's loss of self-control but also suggested an abdication of the strict bourgeois obligations of personal responsibility and self-direction. Authors, playwrights, and artists repeatedly invited audiences to gaze into these alcoholic nightmares, step outside the psychic and physical constraints of respectability, and walk in the dark shadows of the mind.

Whereas the self-made man produced himself through the rational exercise of will, guided by his strength of moral character, the drunkard succumbed to diseased imaginings that sprang from his moral weakness.

While the cult of the self-made man was thus predicated on unquestioning faith in the possibility of self-making and personal transformation, representations of delirium tremens in popular culture spoke to a powerful desire for transcendent experience that exceeded this market-driven identity: to be transformed by the power of imagination rather than to vigilantly constrain and control its impulses through force of masculine will. Narratives of pathological drinking performed this romantic longing for release from the imperatives of the new middle-class ethos: to abdicate the responsibilities inherent in a selfhood determined by individual effort and escape the hardships of an unpredictable boom-and-bust economy in which the common experiences of failure and bankruptcy were understood to result from personal failings. In representations of delirium tremens, inebriates became objects of horror because they suggested that intemperate impulses lay dormant in all individuals, always threatening to erupt from the mind's shadows and wreak terrible destruction on self and family. Yet they also became objects of desire. Inebriates existed outside the rigors of the capitalist market; as they sank into enslavement, they also escaped the requirements of self-control and self-direction, becoming subjects of their own imaginative demons.

The Manny Pokers

In the nineteenth-century cultural imagination, medical descriptions of delirium tremens competed with other ways of understanding alcoholic disease and insanity that derived from common experience. In one sense, delirium tremens was, and is, an intellectual construct—a way of delimiting and understanding a set of characteristic symptoms that can result from heavy drinking. But it is not the only way to understand those symptoms. The condition delirium tremens describes predated the pathological term, and the boundaries of the diagnosis itself narrowed over time, as twentieth-century physicians distinguished delirium tremens from a more general condition they called alcohol withdrawal. Medical literature contains evidence that heavy drinkers were well aware of the medical risks of suddenly abstaining from liquor long before the existence of the delirium tremens diagnosis. A century before Thomas Sutton's first use of the term, George Cheyne wrote, "Nothing is more ridiculous than the common *Plea* for continuing in drinking on, large *Quantities* of *Spirituous Liquors*; *viz.*

Because they have been accustomed so to do, and they think it *dangerous* to leave it off, all of a sudden." Nineteenth-century doctors also commonly complained that patients refused to stop drinking heavily because they feared the health consequences.[5]

In medical wards, physicians, attendants, and patients often used different terms to refer to alcoholic insanity. Physicians in the seaports of Philadelphia, Baltimore, and Boston reported that delirium tremens was "vulgarly" known as "the horrors" among sailors, the poor, and almshouse denizens.[6] In one published case history in 1830, the physician author asserted that the commonly used medical term "mania a potu" was inaccurate. He argued that "vigilant delirium" was more appropriate, but in his hospital ward, there were other names. "A young woman," he wrote, "was brought at night to the institution in that state of temulence with terror called by the nurses and old inmates of the house, the horrors."[7]

To what extent popular experience influenced the thinking and practice of university-trained physicians is difficult to measure, since they had professional incentives to differentiate their knowledge from common beliefs. The horrors do seem to have influenced Benjamin H. Coates, who wrote in his influential essay on delirium tremens that the condition "is well known in the port of Philadelphia [and] is common for sailors."[8] He went on to describe the treatment strategies sailors used for "the horrors." As evidence for his theory that delirium tremens derived from sudden abstinence from alcohol, Coates noted that the horrors were "common for sailors, on first leaving the scene of their frolics for a new voyage."[9]

Popular and medical terms intermingled in midcentury print. The horrors appeared, for instance, in sailor fiction and sailors' autobiographies, which were especially popular in the 1830s.[10] In his memoir published in 1839, the onetime sailor Horace Lane reported that on returning to the United States after nine years at sea, he embarked on a two-week drinking spree in New York City. Having drunk away all his money, he secured a position on an American naval vessel, but "When I had got on board, I was as gone crazy for three days, with the horrors. I felt as if I was beset with fiends of hell, within and all about."[11] In his story papers and novels, Philadelphia author George Lippard wrote many passages portraying street toughs casually referring to delirium tremens as "the tremens," "the horrors," or most often in a humorous and flippant way as "the man with the red-hot poker," a phrase first published in 1842 in T. S. Arthur's *Six Nights with the Washingtonians*.[12] In *Memoirs of a Preacher, or The Mysteries of the Pulpit* (1849), Lippard oddly undermines a tragic scene with humor in a way that suggests his audience's easy familiarity with "the tremens." When a loving wife realizes her husband is suffering from delirium tremens, her innocence prevents her from correctly pronouncing its name:

> Nancy muttered as she caught a glimpse of her husband's fiery eyes and distorted face, "It is the delirious tremens"—poor Nancy! Her knowledge of words was not altogether perfect—"or, it's the manny poker!" These exclamations will no doubt provoke laughter, but had you seen the terrified face of the wife—or heard the accent in which she spoke—laughter would be the last thing in your thoughts. . . . Her husband . . . was now attacked by that fearful avenger of violated nature—*delirium tremens*.[13]

In fleeting references such as these, Lippard uses the "pokers" to render a gritty view of Philadelphia's underworld.[14]

In popular literature, descriptions of delirium tremens most often centered on hallucinations. The disease was a key vehicle not just for popularizing the romantic understanding of apparitions as deriving from diseased or extreme mental states, but for the idea that hallucinations could embody depraved pathological desires. One piece of anecdotal evidence demonstrating this influence can be found in the margins of the Library Company of Philadelphia's copy of the novel *Tom Cringle's Log* (1833). The novel recounts a sailor's adventures in the Caribbean. After indulging in a little rum and wine, Cringle becomes increasingly paranoid and sees an apparition of his dead friend's face appear repeatedly in the folds of his captain's cloak: "The false impression was so strong as to jar my nerves, and make me shudder with horror. I knew there was no such thing, as well as Macbeth, but nevertheless it was with an indescribable feeling of curiosity, dashed with awe."[15] With trembling hands and growing panic, Cringle spends the following days suffering from frightening hallucinations. He later attributes the delusions to the onset of yellow fever, but the passage suggested another disease to at least one nineteenth-century reader. Most likely written about 1840, a penciled note reads, "Mania Potu. If the Author of this is living he is a drunkard ten to one!"[16] The combination of paranoia, trembling hands, and hallucinations linked with heavy drinking made for an easy diagnosis for this middle-class reader.

Growing popular awareness of delirium tremens transformed how readers interpreted literary representations of apparitions and even works that predated the diagnosis itself. The remarkable popularity of Robert Burns's poem *Tam o' Shanter* illustrates this nineteenth-century transformation. Published in 1790, over two decades before the first medical description of delirium tremens, the poem described a habitual drinker visited by demonic creatures, but it did not describe the physical symptoms associated with the disease, such as the distinctive physical tremors.[17] Mid-nineteenth-century readers read the poem in a different way than Burns could have intended. In 1861, for instance, in a dissertation on delirium tremens at the University of Pennsylvania, a medical student cited *Tam o' Shanter* when describing the symptoms of the disease. The student wrote that Burns "has made his hero so plainly express" the horrors of delirium tremens, "that I may be excused for making a quotation."[18]

The demons that haunt Tam are not symptoms of a pathological condition, however. The poem relates how, having again gotten drunk with his cronies, Tam sets out for home on a dark and stormy night. As he passes a lonely church, he sees through a window that demons and witches are dancing inside. The Penn medical student chose Burns's description of this scene to illustrate the horrors of delirium tremens in his dissertation:

> Coffins stood round, like open presses,
> That shaw'd the Dead in their last dresses;
> And (by some devilish cantraip sleight)
> Each in its cauld hand held a light,
> By which heroic Tam was able
> To note upon the haly table,
> A murderer's banes, in gibbet-airns;
> Twa span-lang, wee, unchristened bairns;
> A thief, new-cutted frae a rape,
> Wi' his last gasp his gab did gape;
> Five tomahawks, wi' blude red-rusted:
> Five scymitars, wi' murder crusted;
> A garter, which a babe had strangled;
> A knife, a father's throat had mangled.
> Whom his ain son o' life bereft,
> The grey hairs yet stack to the heft;
> Wi' mair o' horrible and awfu',
> Which even to name wad be unlawfu'.[19]

The beautiful young witch dancing in a "cutty-sark" (a short dress or skirt) entranced the intoxicated Tam. Unable to control his excitement, Tam yells, "Weel done, Cutty-sark!" alerting the supernatural horde to his presence. Tam escapes by fleeing across a nearby bridge on his trusty mare Maggie.

Devoid of psychological terror, *Tam o' Shanter* lacks the frightening am-

Figure 11. Magic lantern slide depicting what Tam saw by peering through the church window. Joseph B. Beale, manufactured ca. late 1860s by Briggs Company. Image courtesy of the American Magic-Lantern Theater.

biguity inherent in a piece like *The Rum Maniac* or the medical narrative of self-castration in which alcoholic demons well up from the maniac's diseased imagination. Tam narrowly eludes the witch's grasp, but he is never in danger of losing his sanity, let alone facing bankruptcy and a squalid death in the almshouse. Tam can escape because the demons have a physical existence outside the operations of his own mind.[20]

Celebrating an unrepentant drunkard subject to demonic and mildly pornographic hallucinations, *Tam o' Shanter* enjoyed enormous popularity among Philadelphia's middle classes at the same moment the temperance movement was thriving and deaths attributed to delirium tremens were at historically high levels in the city. In 1834, audiences paid to see life-size stone statues of Tam, his wife, and two of his drinking cronies in an exhibition at the Masonic Hall.[21] Coincident with the sculpture exhibition,

Figure 12. Magic-lantern slide of Tam's narrow escape.
Joseph B. Beale, manufactured ca. late 1860s by Briggs Company.
Image courtesy of the American Magic-Lantern Theater.

the fashionable Chestnut Street Theatre produced an original production of *Tam o' Shanter*.[22]

Magic-lantern slides produced in the 1860s from engravings first published in Scotland suggest something of Tam's attraction and, more broadly, the popular allure of delirium tremens (see figures 11 and 12). The first slide invites the audience to join Tam as voyeurs, transgressing moral boundaries by peering through a church window to ogle a beautiful, half-naked witch. But the slide also draws on the popular fascination with alcoholic hallucinations. The other supernatural creatures are also objects of voyeurism as Tam wonders at their depraved celebration. The scene of riotous dancing dissolves into horror in the second slide as Tam flees the demons his intoxicated sexual desire has angered. In the darkened theater, midcentury audience members peered into the alcoholic imagination, glimpsed the debauchery of witches and warlocks, and joined Tam in his thrilling escape. The quaint Scottish verse and Burns's respectable literary reputation allowed middle-class audiences to modestly indulge and also to triumph over the frightening demons and prurient desires lurking within their own minds.

Demonizing Rum

Feeding this popular fascination, midcentury authors drew on supernatural symbology to explore the frightening power of pathological drinking. Perhaps more than any antebellum American writer, Edgar Allan Poe illuminated the literary potential of intoxicated states of consciousness. Living and writing in Philadelphia in the 1830s and 1840s, Poe used temperance imagery in many of his stories, constructing individualistic explorations of the psychology of perversity that draw on contemporary theories of the mind to produce unsettling effects. One of his critics explained that Poe exploited "those strange unsolved phenomena in the human mind, which the terms mesmerism and somnambulism serve rather to disguise than to discover, and sweat out from their native soil superstitions far more powerful than those of the past."[23] Stories such as "The Black Cat," "King Pest," and "The Cask of Amontillado," among others, associated intemperance with fantastic imagery, the supernatural, and radical evil. In "The Black Cat," for instance, Poe describes alcohol as a demonic "disease." In the story, the "instrumentality of the Fiend Intemperance" destroys the domestic happiness of a man who then murders his wife with an ax.

Like *Tam o' Shanter*, Poe's first published short story, "Metzengerstein" (1832), centered on a horse and an intemperate rider, but in Poe's story the rider's self-destructive compulsion constituted the principal drama. Imagery and themes from temperance literature create a powerful symbolic struggle between man and beast. The central premise, for instance, is metempsychosis, a metaphor frequently used in temperance literature. Metempsychosis refers to the transmigration of the soul after death from one human or animal into another. Temperance advocates cited metempsychosis when describing the effects of intoxication. In his famous pamphlet *An Inquiry into the Effects of Spirituous Liquors*, Benjamin Rush wrote that the ancient Greek doctrine of metempsychosis "was probably intended only to convey a lively idea of the changes which are induced in the body and mind of many by a fit of drunkenness. In folly, it causes him to resemble a calf,—in stupidity, an ass—in roaring, a mad bull,—in quarreling, and fighting, a dog,—in cruelty, a tiger,—in fetor, a skunk,—in filthiness, a hog,—and in obscenity, a he-goat."[24] "Metzengerstein" uses allusions to delirium tremens, metempsychosis, intemperance, and irrational compulsion to create potent and ambiguous imagery.[25]

For Poe, intemperance and death were not just abstract literary subjects, and "Metzengerstein" may have been inspired by a personal tragedy. The story was published just five months after Edgar's beloved older brother, Henry, died of what a family friend called "intemperance." Henry was just twenty-four. It is entirely possible that he suffered delirium tremens and that Edgar witnessed his affliction. When used to designate cause of death, "intemperance" was an ambiguous term. It commonly referred to delirium tremens, but it also meant alcohol poisoning or some other disease perceived to be brought on by heavy drinking. Henry died in Baltimore on August 1, 1831, and the funeral was held in the home where Edgar was living. Henry's death had a profound impact on his younger brother. As orphans, Henry and Edgar had "clung together psychologically as to be nearly one person."[26] Henry was also an author, and he and Edgar collaborated on romantic poetry. In 1831, after years of separation, Edgar and Henry had reunited in Baltimore, but Edgar described his brother as "entirely given up to drink & unable to help himself."[27] The story of an orphaned nobleman given to debauchery, "Metzengerstein" appeared in Philadelphia's *Saturday Courier* on January 14, 1832. Edgar likely wrote the story in the months after Henry's death.[28]

Whereas Maggie carries Tam safely away from the "ghaists and houlets," Poe's story contemplates a young aristocrat locked in a death struggle with a demonic black horse. Poe sprinkles the story with allusions to the man's mental state, describing him variously as suffering from "mania," "anxiety," and an "unnatural fervor." The story begins when, after four days of "shameless debaucheries," the young and recently orphaned Baron Metzengerstein succumbed to paranoia and hallucination. Brooding in his palace, he stared at an ancient tapestry that pictured "an enormous and unnaturally colored horse." The image gripped his attention like a "spell," and "he could by no means account for the overwhelming anxiety which appeared falling like a pall upon his senses." Suddenly, "to his extreme horror," the image of the horse seemed to emerge from the tapestry: "The neck of the animal . . . extended, at full length, in the direction of the Baron. The eyes, before invisible, now wore an energetic and human expression, while they gleamed with a fiery and unusual red; and the distended lips of the apparently enraged horse left in full view his sepulchral and disgusting teeth."

Terrified, the Baron flees to his castle courtyard and finds that his stable hands have just captured a magnificent black horse that had been wandering free in the castle grounds. Poe suggests that this horse's soul is that of the Baron's hereditary enemy, Duke Berlifitzing, who had died that night in a stable fire that the Baron set during his destructive debauchery.[29] The black steed thus appears ambiguously as both natural and supernatural. Is it a demon spawned in the fiery death of Berlifitzing, its appearance augured in Metzengerstein's debauched hallucination? Or is it simply a magnificent horse? Poe never resolves this ambiguity.

The horse embodies the mysterious impulses that drive the Baron to insanity, horror, and death. "The Baron's perverse attachment" to the horse "became in the eyes of all reasonable men, a hideous and unnatural fervor." Intoxicated by the power of the "ferocious and demon-like" animal, the Baron obsessively rides the horse day and night, desperately seeking to control the evil beast. But the horse has its own dark purpose. Observers noted "an unearthly and portentous character to the mania of the rider."[30] Desire, loathing, and ecstasy mingle in the Baron's obsession. One of his servants, "whose opinions were of the least possible importance," Poe tells us, "had the effrontery to assert that his master never vaulted into the saddle, without an unaccountable and almost imperceptible shudder," but that "upon his return from every habitual ride . . . an expression of triumphant malignity distorted every muscle in his countenance."[31] In a perverse recasting of Tam's ride, the homoerotic image of the Baron day and night being "riveted to the saddle" of his reincarnated hereditary enemy heightens the depraved nature of the struggle for dominance between man and demon.[32]

The Baron's ultimate incineration in a preternatural fire evoked contemporary temperance accounts of the spontaneous combustion of drunkards. One typical night, the "Baron descended, like a maniac from his bed chamber," mounted the horse, and galloped off into the forest. During his ride, however, "a livid mass of ungovernable fire" engulfed his castle. Those gathered to watch were amazed to see the phantom steed galloping out of the forest, carrying the Baron toward his doom. Horrified witnesses reported that "the career of the horseman was indisputably . . . uncontrollable." The Baron was carried toward the fire stricken with fear: "The agony of his countenance, the convulsive struggling of his frame, gave evidence of superhuman exertion: but no sound, save a solitary shriek, escaped from his lacerated lips, which were bitten through in the intensity of terror."[33] As the steed charged into the burning palace, "its rider, disappeared amid the whirlwind of chaotic fire":

> The fury of the tempest immediately died away, and a dead calm sullenly succeeded. A white flame still enveloped the building like a shroud, and, streaming far away into the quiet atmosphere, shot forth a glare of preternatural light; while a cloud of smoke settled heavily over the battlements in the distinct colossal figure of—*a horse*.

The Baron's demise is a spectacle. Far from eliciting pity or revulsion, Poe portrays the final, desperate struggle as inspiring awe among the witnesses. Poe casts the inebriate as a melancholic Hamlet, consumed by the phantom of his diseased imagination.

Poe's writing thus expressed a broader development in the history of the supernatural. Authors increasingly represented phantasms not as simply hallucinations linked with diseased or altered states of mind—like mania, fever, or opium dreams—but as embodying irrational, compulsive, and immoral impulses such as the drunkard's craving for drink. The growing prevalence of these images in popular culture had a social basis, also exemplified by Poe. Stories of alcoholic demons thrived during a period when Americans were increasingly subject to new economic relationships and unseen market forces. The power of liquor to paralyze the will and drive the careless drinker into penury was powerfully alluring at a moment when the fortunes of middle-class entrepreneurs, farmers, artisans, and mechanics became bound up in far-flung networks of exchange, new forms of credit, the ebb and flow of paper currency, fluctuating land values, and an unpredictable economy.[34] At no time were these forces more apparent and threatening than during the economic depressions that followed the financial panics in 1819 and 1837. These downturns threw many entrepreneurs into bankruptcy and subjected previously independent artisans and mechanics to new forms of capitalist wage labor. While antebellum culture defined success and failure as commensurate with an individual's character and strength of will, impersonal market forces drove even the respectable and hardworking into dependence.

The cultural understanding of liquor's power to unman the drunkard, to rob him of his will, thus expressed the market's power to deprive men of their economic independence. Published in May 1841 while Poe was living in Philadelphia, the short story "A Descent into the Maelström" centered on a temperance metaphor from the 1830s and 1840s. In 1833, the editors of a Universalist magazine in Boston wrote,

> No man is safe who drinks cautiously. If you get into the habit at all, it is like the rash or ignorant mariner entertaining the disk of the Maelstrom, that

> great whirlpool on the coast of Norway. He cannot keep upon the edge. Each circumstance carries him nearer to the centre, and of course to irretrievable destruction. His only safety lies in keeping out of the current, and at a distance from it. Some people drink to drown sorrow. . . . They are about as reasonable as the mad commander of a vessel, who . . . ran his vessel into the whirlpool, that the dizziness produced by the rapidity with which he made the circumference, might cause him to forget his troubles.[35]

Other writers used the same metaphor to dramatize liquor's mysterious power to suck the unvigilant into darkness. In the temperance novel *The Glass*, Maria Lamas lamented the "young men one by one caught in the whirl of that great Maelstrom, more fatal than its Norwegian prototype, and lost forever to society, to themselves and to heaven."[36] Poe published his story in the midst of the proliferation of Washingtonian societies, which coincided with the economic depression following the Panic of 1837. "A Descent into the Maelström" alludes to the Washingtonian "conversational meetings," especially their function as mutual benefit societies. Small-scale artisans, mechanics, and manual workers relied on these organizations to help them cope with the hard economic times.[37]

In Poe's rendering, the whirlpool evokes both the dire consequences of economic failure and the frightening depths of the human imagination. Evoking the new style of temperance meetings, "A Descent into the Maelström" recounts a conversation between two men: a fisherman and Poe's narrator. From their vantage on the Norwegian coast, the two gaze out over the ocean to witness the terrifying power of the maelstrom. The narrator relates in detail the awesome dimensions of the natural phenomenon, which forms, diminishes, and disappears according to a very regular schedule. As the maelstrom achieves its full shape, the narrator is overawed:

> In a circle of more than a mile in diameter. The edge of the whirl was represented by a broad belt of gleaming spray; but no particle of this slipped into the mouth of the terrific funnel, whose interior, as far as the eye could fathom it, was a smooth, shining, and jet-black wall of water . . . speeding dizzily round and round . . . and sending forth to the winds an appalling voice, half shriek, half roar, such as not even the mighty cataract of Niagara ever lifts up in its agony to Heaven.[38]

Trees, whales, bears, boulders, and ships of all sizes swirled into the whirlpool's unfathomable inky depths. The fisherman and narrator agree that scientific explanations for the existence of the maelstrom were "altogether unintelligible, and even absurd, amid the thunder of the abyss."[39]

The fisherman then relates a tale that in its outline closely resembled the stories working people brought to the conversational meetings of temperance beneficial societies. The fisherman and his two brothers had owned a small vessel. While most boats made the long journey to safe but only modestly productive fishing grounds, he and his brothers chose to fish close to the maelstrom because of the abundance and higher quality of the fish there. Writing while the memory of the Panic of 1837 was still fresh, Poe casts their choice in the language of capitalist finance: "In fact, we made it a matter of desperate speculation—the risk of life standing instead of labor, and courage answering for capital."[40] One day their luck turned bad. Returning home after a productive day's fishing, the boat was caught in a large storm. The winds immediately blew one brother overboard and then drove the boat into the maelstrom, which had just reached its peak fury. As they approached the lip, the fisherman's remaining brother was clinging to a water cask, a potent symbol of temperance. Panicked, the brother abandoned the cask and frantically grabbed at a handhold on the mast that the fisherman had been using. The fisherman realized his brother had gone insane: "I never felt deeper grief than when I saw him . . . a raving maniac through sheer fright."[41]

The fisherman maintained his sanity only by surrendering himself to the maelstrom. Relinquishing the handhold, he made his way astern and grasped the water cask his brother had abandoned. Meanwhile, the boat slipped over the lip and into the abyss. Rather than plummeting down, it hung on the black wall of water: "Never shall I forget the sensations of awe, horror, and admiration with which I gazed about me. The boat appeared to be hanging, as if by magic, midway down, upon the interior surface of a funnel prodigious in circumference, immeasurable in depth."[42]

Compelled to gaze downward into the pit itself, he searched its depths but reported that "still I could make out nothing distinctly, on account of a thick mist . . . over which there hung a magnificent rainbow, like that narrow and tottering bridge which Mussulmen [Muslims] say is the only pathway between Time and Eternity."[43] Abandoning any effort to steer the ship that constituted his economic livelihood, the fisherman can contemplate the infinite.

The story's ending reflected the changing perceptions of the inebriate in the 1840s. As the new conversational meetings made reforming the drunkard and rum seller their central goal, so the fisherman avoided the fate of the Baron Metzengerstein. Remembering past geometry lessons, he realized the water cask he clung to might help him float on the side of the maelstrom until its motion diminished. Forced to abandon his maniacal brother, he lashed himself to the barrel, threw himself overboard, and floated to salvation.

Poe thus expressed the perilous position of many small-scale artisans and entrepreneurs in 1841. In the context of economic collapse, "A Descent into the Maelström" illuminates the allure of the dark imagination as a new foundation for meaning and identity that exceeds the imperatives of the capricious market economy. The first brother is immediately blown overboard. The second brother goes insane because he rejects the water cask as a potential escape and clings to the vessel of their "desperate speculation," even as the whirlpool destroys it. Abandoning his ship to float on the dizzying whirl, the fisherman gains knowledge accessible not through the force of masculine will and industry, but through surrender to the wild, uncontrolled nature of the soul. Like a holy man from the East, he returns from beyond to share his wisdom. Rejecting the pursuit of gain enables him to look beyond earthly existence, although the knowledge he gains comes at a steep price. Like any drunkard, his journey has left him broken and old beyond his years. He is initially speechless from the horror of his experience, and his rescue by a passing boat again evokes a local temperance society holding a conversational meeting: "Those who drew me on board were my old mates and daily companions—but they knew me no more than they would have known a traveler from the spirit-land."[44] While the Baron Metzengerstein meets his demise clinging to the back of his alcoholic demon, the fisherman survives by abandoning ship and radically reconstituting himself as a shaman.

Poe suggests that one can cheat the dizzying attraction of the maelstrom to briefly glimpse a realm of imagination and return to tell the tale, even if the price is health and well-being. His casting the temperance narrative as a mystical journey through a "spirit land" also seems an apt description of a magic-lantern slide depicting the horrors of delirium tremens, produced no later than 1859 (see fig. 13). Used in temperance lectures, this slide was part of a series showing the progressive degradation of the drunkard. In this image, a bleak and forbidding landscape provides the backdrop for the deranged man, in stocking feet, poised atop a crag. The size and diversity of the supernatural horde speaks to the depths of horror and despair that liquor threatens to unleash in the mind.

Audiences viewing this slide knew the demons were imaginary. What

Figure 13. Magic-lantern slide depicting the horrors of delirium tremens. Joseph B. Beale, manufactured ca. 1859 by the T. H. McAllister Company, New York. Image courtesy of the American Magic-Lantern Theater.

makes the image so frightening and fascinating is the realistic portrayal of the man suffering hallucinations. Unlike Tam eluding his supernatural pursuers, this man cannot escape the legions of horrid creatures welling up from the depths of his own mind. Cultural representations of delirium tremens thus constructed compulsive drinking as a psychological struggle with imaginary demons. Here the technology of the magic lantern gave audience members a voyeuristic peek into the dark shadows of the alcoholic mind—a fleeting glimpse into the maelstrom. Safe in the knowledge that the image was an exercise in moral instruction, audience members could contemplate the drunkard as a figure existing outside the strictures of the will, a subject of the imagination. What the drunkard could not escape, audience members longed to see.

A Disease of Social Decay

Because of its close association with bankruptcy, business failure, and social downfall, delirium tremens also became a likely subject for authors across the political spectrum commenting on the moral dimensions of capitalism, wealth, and social inequality. Perhaps because of the high visibility of a wealthy medical faculty and the large medical student population, Philadelphia writers especially used the disease to expose the city's increasingly rigid class structure. Some commentators drew on delirium tremens in radical condemnation of the social pretensions of Philadelphia's middle class. For these social critics, the disease's symbolic power derived from its subversive implications. But in drawing on the disease for dramatic effect, these authors paradoxically reinforced moral stigmas that associated intemperance with depravity and evil, legitimating disparities of wealth and class.

In general, literary and theatrical portrayals of delirium tremens reflected the social makeup of the disease in the city's hospitals and institutions. Of antebellum Philadelphians who died of delirium tremens, 85 percent were male, and most were middle-aged and white.[45] In literature and theater, the victims of delirium tremens were also overwhelmingly white men, with only a few exceptions. In Augustine Duganne's novel *The Knights of the Seal, or The Mysteries of the Three Cities* (1848), for instance, a Spanish pirate who had been trying to conceal his true identity in Philadelphia society contracts delirium tremens after committing a murder.[46] In Maria Lamas's temperance novel *The Glass, or The Trials of Helen More* (1849), a woman corrupted by her desire to be wealthy and fashionable suffers a three-day attack of delirium tremens during which her son dies while locked in the attic.[47] But even these exceptions retained the disease's link with social status: the corruption inherent in ill-gotten gains and the dangers of fashion-conscious women corrupting the bourgeois home. Portrayals of the disease striking the city's poorest and most marginal groups, most notably the substantial black population, were extremely rare.

Hallucinations were thought to derive from an overactive imagination, which implied a rich inner mental life and thus a person of some education and respectability. Members of the new middle class emerging in the 1830s distinguished themselves from other social groups not only through morals and habits, but also through behavior and appearance. Etiquette books, magazines, and sentimental literature popularized modes of physical conduct rooted in eighteenth-century notions of gentility.[48] Middle-class ideals demanded strict physical and emotional self-restraint. Manuals detailed rules for posture, arm movements, walking style, hand gestures, and facial expressions. Spitting, coughing, yawning, and other bodily processes were to be concealed or were forbidden entirely. These outward behaviors demonstrated inner moral worth, thus legitimating the middle-class aspirant's place in society.[49] The drunkard suffering from delirium tremens violated all these modes of behavior.

These meanings are evident in the novel *Sheppard Lee* (1836), written by the physician Robert Montgomery Bird.[50] Trained at the University of Pennsylvania in the 1820s, when medical interest in delirium tremens was at its peak, Bird drew on his professional experience in using the disease to criticize the unrestrained greed he saw as prevalent in antebellum society. In the novel, the wealthy miser Abram Skinner confronts his youngest son Abbot. Hateful, profligate, bankrupt, and dissolute, Abbot is also in dire need of his father's financial support. He had accrued a $20,000 debt through drinking and gambling. The abstemious Abram offers to pay the debt if Abbot will stop drinking. The son replies, "You have asked the question a month too late. Look," and shows his father that his hands are displaying the distinctive tremor associated with delirium tremens. Abbot asks with a morbid laugh, "Do you know what that means? . . . It means . . . that death is coming."[51] To quit drinking now, he is saying, will only aggravate the disease and hasten his inevitable demise.

Abbot's disease expresses his intemperate habits, his immoral pursuit of gain through gambling, and his father's failure to provide proper parental guidance, no doubt because of Abram's miserly nature. A month after this confrontation, Abbot attempts to murder his father but is deathly ill and unable to accomplish his goal. Abram looks on in horror as his son succumbs to delirium tremens:

> He lay, at times, the picture of terror, gazing upon the walls, along which, in his imagination, crept myriads of loathsome reptiles, with now some frightful monster, and now a fire-lipped demon, stealing out of the shadows and preparing to dart upon him as their prey. Now he would whine and weep, as if asking forgiveness for some act of wrong done to the being man is most constant to wrong—the loving, the feeble, the confiding; and anon, seized by a tempest of passion, the cause of which could only be imagined, he would start up, fight, foam at the mouth, and fall back in convulsions. Once he sat up in bed, and, looking like a corpse, began to sing a bacchanalian song; on another occasion, after lying for many minutes in apparent stupefaction, he leaped out of bed before he could be prevented, and, uttering

> a yell that was heard in the street, endeavoured to throw himself from the window.[52]

Significantly, Abbot performs the disease while his father watches. For Bird, the spectacle of delirium tremens teaches the miser the true consequences of immoral gain. Bird writes that Abbot's "last raving act of all was the most horrid." Abbot hallucinates that he is murdering his father and dies with a horrid chuckling laugh, "exulting, in imagination, over a successful parricide."[53]

In criticizing immoral wealth, *Sheppard Lee* affirms moral pursuit of gain. The social meanings of delirium tremens thus functioned here much as they did in medical literature: to rationalize the failure of respectable individuals and reaffirm emerging middle-class values celebrating temperate habits, moderation, and delayed gratification. Abbot's disease begins with a loss of control over his trembling hands and ends with the loss of all physical and emotional restraint. Narratives of delirium tremens generally dwelled on patients' countenances, which became contorted with horror. The loss of physical control expressed the loss of inner control as the will was overwhelmed by the unbridled imagination. Abbot's terror derives from the demons and devils that "steal out of the shadows." In his habits, inner thoughts, and outward behaviors, the drunkard suffering delirium tremens became the antithesis of the middle-class ideal of the self-made and self-controlled.

The class expectations of an elite profession initially shaped the meanings associated with delirium tremens, but writers and social commentators drew on the disease for their own purposes. Some of the earliest mentions of delirium tremens to be published outside medical publications appeared in the mouthpiece of Philadelphia's early labor movement, the *Mechanic's Free Press*. The author wrote under two pseudonyms, Peter Simple and the Night Hawk. In poetry and short fiction that often criticized the moral pretensions of the wealthy and respectable, he portrayed delirium tremens as potentially liberating. In one short story of the social downfall of a well-to-do but naive young man, Peter Simple equated the madness of mania a potu with the ecstatic mental states of opium dreams as described by Thomas De Quincey in his *Confessions of an English Opium-Eater* (1821). "The Confessions of a Drunkard" (1829) was the first-person narrative of a promising young man ruined by liquor. He wasted his inherited estate on drinking and gambling, murdered a friend in a fit of rage, and seduced an innocent young woman before becoming a victim of delirium tremens. For Peter Simple, delirium tremens epitomized the youth's depravity, yet the disease also allowed him a more authentic vision of human existence. Describing the experience of lying on a city street suffering wild delusions, the youth confessed,

> Yet a secret pleasure danced around my heart even there. I regarded not the scoffs of the rich, as they rolled past in their carriages, whose wheels passed within an inch of my feet—they could crush, but could not subdue. I lay basking in the sunshine of desperation—the world had rolled from me in dense clouds; her beauty was like the rainbow, momentary to my soul. . . . I shouted in the triumph of madness over reason—blasphemed in the wildness of my dream, and sank into total forgetfulness amid the hum of voices and the rush of carriages. . . . Oh! That such a fire could consume us at once—burn out this brief existence clogged with artificial matter to carry us through the world.[54]

Simple recognized in Philadelphia physicians' descriptions of mania a potu the same sense of desire that pervaded the fascination with opium dreams apparent in De Quincy in particular and romanticism more generally, but Simple turned it into a vehicle for social criticism. Like opium dreams, delirium tremens was a personal transformation that enabled the youth to step outside class expectations and explore the potential of the diseased imagination—a transcendent dream state—even if he paid in madness and death. In the *Mechanic's Free Press*, the youth's declaration that the carriages of the wealthy could "crush, but could not subdue" articulated the defiant anger of the early labor movement.

Even if some authors used delirium tremens as a vehicle for social criticism, the visibility of the disease in popular culture nevertheless reflected conservative notions of social difference, which stigmatized poverty as a moral failing. Narratives exploring the psychology of the diseased inebriate performed the cultural work of distinguishing between the intemperate and depraved poor and those previously respectable individuals who had fallen into poverty and disgrace. In literary accounts as well as in Philadelphia's medical wards, the disease did not strike middle-class people exclusively but rather attacked any capable individual who failed to live up to the ideals of thrift, industry, and self-restraint demanded by middle-class ideals and the rigorous capitalist marketplace. While the unregenerate poor remained so because they lacked the will to lift themselves out of suffering and depravity, diseased inebriates fell victim to their powerful imaginations, a creative faculty the poor lacked altogether.

In representations of delirium tremens and pathological drinking, these distinctions became even more pronounced by the 1840s. Writing amid the economic wreckage that followed the financial panics of 1837 and 1839, the best-selling author George Lippard saw a city sharply divided between haves and have-nots. Some of Lippard's novels were so wild that commentators remarked that his writing style seemed inspired by delirium tremens. In 1849 the journalist and reformer Jane Grey Swisshelm wrote that "the Lippard style . . . requires the writer to be born with St. Vitus's dance, to be inoculated with the Delirium Tremens, take the nightmare in the natural way, get badly frightened at a collection of snakes, and write under the combined influence of these manifold causes of inspiration."[55] This quotation may be the source of rumors among Philadelphia writers that delirium tremens caused Lippard's death in 1854. The Philadelphia writer Charles D. Gardette told Walt Whitman in 1860 that Lippard was "handsome" and "Byronic" and died "mysteriously, either of suicide or mania a potu."[56] Lippard actually died of tuberculosis, but the rumor demonstrates how the association of delirium tremens with the diseased imagination made it somehow "Byronic."

In Lippard's *Memoirs of a Preacher*, delirium tremens provides the vehicle for a stunning censure of the moral depravity of Philadelphia's ruling class. In the novel, John is an honest tradesman who has been driven to heavy drinking by his sudden inability to find work and his despair over his family's poverty. In a desperate attempt to avoid eviction, John tricks his greedy landlord, Israel Bonus, into coming in contact with a woman who has just died of smallpox. At the same moment that Israel becomes convinced he has contracted the deadly disease, John succumbs to an attack of delirium tremens. In his madness he grabs Israel, who is desperate to flee the house. John relishes torturing Israel, reminding him that smallpox festers in the squalid dwellings that Israel profits from:

> This pestilence takes wife, and children, and rots a liven' bein' afore death, so that his own mother wouldn't know him. Who keers? But, Isr'el, 'spose this disease gets hold o' th' one as planted it—get's hold o' you, and settles up old scores with you, and digs into your heart, as it has dug into the hearts of these miserable vagabonds in your Court—what then?[57]

Shamanlike, John's alcoholic delirium enables him to speak truth to Philadelphia's rapacious ruling classes, whose desperate greed perpetuates the social conditions that create intemperance, disease, and suffering. His disease frees his insight from the mental constraints and requirements of social class.

Lippard offered the horrors of poverty as a context for explaining the intemperance of the poor, and despair at finding no work led to the honest tradesman's attack of delirium tremens. In his best-selling novel *The Quaker City, or The Monks of Monk Hall* (1844), he offered no equivalent explanation for the intemperance of the wealthy. Instead, drunkenness expressed and inspired the radical evil committed by wealthy citizens. While *Quaker City* includes passing references to the "pokers" and delirium tremens, the novel also associates alcohol with hallucinations, paranoia, and dread in ways that suggest the disease. Drunkenness, delirium tremens, opium dreams, chloroform intoxication, mesmeric trances, apocalyptic dreams, spiritual ecstasy, and religious enthusiasm, along with ill-defined species of mania and delirium, create a surreal pastiche of antebellum cultural ideas about the imagination, insanity, disease, and intoxication.

Lippard thus drew on the popular fascination with imaginative phenomena, a fascination especially strong in middle-class culture, to portray city elites as shockingly depraved. *Quaker City* contains five sprawling plot lines, but the main story line opens with an extended alcoholic nightmare. While drunkenly carousing in an oyster cellar, the rake and confidence man Gus Lorrimer involves Byrnewood Arlington, a young businessman, in a plot to deceive, seduce, and deflower the innocent daughter of a wealthy merchant. Lorrimer plans to use a promise of marriage to lure her to "Monk Hall," a secret brothel maintained by the city's wealthiest citizens. Lorrimer leads Arlington to the ancient mansion, which is filled with trapdoors and hidden passages and haunted by phantoms and disembodied screams. Descending a "subterranean stairway, surrounded by the darkness of midnight," Arlington and Lorrimer come upon the monks of Monk Hall gathered in the clubroom. Arlington recognizes the monks as "poets, authors, lawyers, judges, doctors, merchants, gamblers, and . . . one parson."[58]

The entire passage describing the night Arlington spends in Monk Hall intertwines heavy drinking with references to insanity, horror, disease, and the supernatural. Lippard's description of the monks' clubroom, for instance, reads like an alcoholic hallucination that strongly resembles accounts of delirium tremens. Heavily intoxicated, Arlington gazes at huge images of Bacchus and Venus, while the walls are alive with intricate "uncouth sculpturings of fawns and satyrs, and other hideous creations." Presiding over the hall, above the chair of the president, is a skeleton figure draped in heavy black robes:

> His right hand raised on high a goblet of gold. From beneath the shadow of the falling cowl, glared a fleshless skeleton head, with the orbless eye-

> sockets, the cavity of the nose, and the long rows of grinning teeth, turned to a faint and ghastly crimson by the lampbeams. The hand that held the goblet on high, was a grisly skeleton hand; the long and thin fingers of bone, twining firmly around the glittering bowl.[59]

Lorrimer goes off to effect his seduction, leaving Arlington in the clubroom. Eventually alone, as the monks drift away or pass out, Arlington is overcome with "involuntary horror" when he looks up at the skeleton monk and sees "the teeth move in a ghastly smile." "The orbless sockets, gleaming beneath the white brow, flashed with the glance of life, and gazed sneeringly in his face." "How strange I feel!" thinks Arlington, "Can this be the first attack of some terrible disease?" Overcome with dread, he is convinced he is going insane. The hallucinations and paranoia drive him into the perilous hallways of Monk Hall in search of Lorrimer.[60]

Arlington's drunken hallucinations and inchoate paranoia turn into a terrifying nightmare. He finds Lorrimer and discovers that the woman the rake has deceived is actually Arlington's own sister Mary—which surprises Lorrimer as well. Refusing to relinquish the virginal young woman, Lorrimer enters into a plot with the disfigured proprietor of Monk Hall, named Devil-Bug, to lock Arlington in the tower and then murder him. After the distraught Arlington paces the tower room for several hours, Devil-Bug brings him food and a bottle of wine drugged with opium. The mixture suggests laudanum, commonly used to treat delirium tremens. Strengthening this association, Devil-Bug says privately of the wine, "Got it from the doctor, who used to come here—dint kill a man, only makes him mad-like. The Man with th' Poker isn't nothin' to this stuff." Devil-Bug leaves the food for Arlington and places a coal-burning stove in the fireplace. The chimney has been bricked over, however, and once locked, the room is airtight. Between the drugged wine, the fumes from the burning coal, and a secret trapdoor in the floor leading to a three-story fall into Monk Hall's cellars, Arlington's drunken spree has come to an end in a room of certain death.[61]

His experience in the tower articulates a monstrous satire of middle-class values. The room is a library, a potent symbol in midcentury America of middle-class respectability and mental culture.[62] Lippard comments on the falsity of such intellectual pretensions when Devil-Bug reveals that the books in this library are not real but only skillfully painted on the walls. After Devil-Bug leaves, Arlington is unaware that the coal stove is slowly poisoning the air. He is suddenly thrown into "a violent delirium of thought." He gasps, "There is a hand grasping me by the throat—I feel the fingers clutching the veins, with the grasp of a demon!" Realizing his peril, he lunges to extinguish the stove but only falls "to the floor, like a drunken man."[63] Much like the fate of the castrated cigar maker, the young businessman is brutally unmanned as he lies prostrate and listens to his sister screaming in a nearby room as Lorrimer rapes her. Indeed, his own pursuit of drunken pleasure has made him complicit in the act. In one last burst of strength, Arlington drinks the drugged wine, and the stimulation of the opium momentarily awakens him. Lunging forward in an attempt to escape, he trips a concealed spring. The air released by the huge trapdoor momentarily revives him, but too late: "Tottering in the darkness on the very verge of the sunken trapdoor, he made one desperate struggle to preserve his balance, but in vain. For a moment his form swung to and fro, and then his feet slid from under him; and then with a maddening shriek, he fell. 'God save poor Mary!'"[64]

The library of Monk Hall is a theater of horrors, a thinly veiled illusion of middle-class respectability perched precariously above a black pit of shame, despair, and death. In an unstable economy with vast disparities of wealth between rich and poor, Lippard draws on the imagery of intoxication, hallucination, and insanity to deplore the shallow moral and intellectual pretensions of the middle class. The trapdoor highlights the perilous nature of their tenuous social position. Arlington's physical fall into the bowels of Monk Hall dramatizes his moral and social downfall caused by drunken debauchery.

Lippard used allusions to delirium tremens, intoxicated trance states, and temperance imagery to construct an alcoholic nightmare that epitomized the radical evil of the city's elite. But by placing the intemperance of the wealthy in the dark shadows of the imagination and portraying the poor as an intemperate mass, he replicated the rationalizations of economic disparities manufactured by Philadelphia's prominent citizens. Lippard's descriptions of the clubroom in Monk Hall, for instance, romanticized the intemperance of the wealthy as occurring in a nightmare realm of forbidden indulgences, in stark contrast to the intemperance of the poor, festering in squalid and filthy alleys. Both rich and poor drink, but the meaning of their drinking differs. For the poor, drinking derives from and legitimates their social circumstances. For the respectable middle class, intemperance is irrational, mysterious, and hideously fascinating. It emanates from the imagination, a faculty associated with dreams and artistry.

Middle-Class Dream / Alcoholic Nightmare

Lippard both participates in and comments on the perverse fascination with social downfall, which was so evident in delirium tremens narratives and, more generally, the popular culture of the 1840s. In the midst of Arlington's death struggle in the tower, he is startled to see that "a hideous face glared upon him, from the aperture of the bookcase, like some picture of a fiend's visage." The "hideous face, with a single burning eye, with a wide mouth distending in a loathsome grin, with long rows of fang-like teeth" appears as a demonic hallucination, but it is also the face of Devil-Bug, who gleefully watches through a glass panel as Arlington suffers and falls. Placing Devil-Bug's distorted visage in the false library bookcase, Lippard satirizes the voyeurism inherent in representations of young men falling into depravity. Indeed, in this moment, Arlington's hallucination is the leering face of middle-class voyeurism.

In some of the most popular melodramas of the century, plays that catered specifically to the middle class, audiences gawked as actors graphically portrayed the suffering occasioned by delirium tremens. In antebellum America, commercial theaters commonly presented plays for one night only, seeking to draw audiences with new programs each night. First staged in Boston in 1844, in its first year *The Drunkard* was performed in that city more than 140 times. Recognizing its tremendous commercial potential, P. T. Barnum brought the play to his American Museum in New York City. Barnum presented *The Drunkard* on consecutive nights for years, providing a new business model for commercial theaters.[65] Traveling productions of *The Drunkard* also toured nationally through the end of the century. The anonymous author attributed the play's "triumphant success" in Boston to its "terribly real" portrayal of human frailty, but "particularly the scene of *delirium tremens*, . . . which though far short of the horrors of that dreadful malady, and appearing, to those unacquainted with the disease to be overstepping the bounds of nature, was true to the letter, and universally acknowledged to be the most natural, effective acting ever seen in this city."[66]

Written by a friend of Poe's, T. S. Arthur's *Ten Nights in a Barroom* (1854) was originally a best-selling novel. The story enjoyed even more popularity as a play. First performed in 1858, the audience for the stage adaptation grew in the years after the Civil War, enduring on stages nationwide until well into the twentieth century. The play "appeared practically everywhere . . . in churches, in tents, in temperance and town halls, in opera houses and on showboats."[67] Manufacturers mass-produced several adaptations of the play for the magic lantern, which enabled temperance activists and theatrical entertainers to present the story in myriad settings.[68]

Becoming major events in nineteenth-century American culture, *The Drunkard* and *Ten Nights in a Barroom* popularized a new theatricality of pathological drinking. They drew on the Washingtonian narrative of the reformed drunkard, conflating delirium tremens with several popular forms of theatrical entertainment, joining elite medical discourse with mass culture. Mesmerism and the phantasmagoria were both forms of entertainment based on the new science of the mind that emerged in the late eighteenth and early nineteenth centuries. Theatrical performances of delirium tremens drew on the imagery of mesmerism and the supernatural to dramatize irrational impulses, depraved desires, and demonic imaginings. In *The Drunkard* and *Ten Nights in a Barroom*, nationwide audiences learned what delirium tremens looked like and, in the process, learned that casual tippling could become a physiological and psychological compulsion of frightening power.

These dramatic portrayals of delirium tremens derived in part from the enduring popularity of the phantasmagoria. Visual images were especially significant in theatrical performances of the disease, as psychological impulses became embodied in phantoms and other illusions. By the 1840s, hallucinations had long been a staple of antebellum entertainment. Because the supernatural was the primary subject matter of the early shows, the lantern used in phantasmagoria performances became known as the "magic lantern," even as subject matter diversified to include images of classical paintings, European cities, Chinese artwork, popular fiction, physiology, significant historical figures, and images of the Civil War. As the technology developed, promoters promised "Chromatotropical, Phantasmagorical, and Dioramic views," "Dissolving Views!" "MOVING Astronomical Diagrams," and a "fast young gent" riding a horse, often all in one show.[69] The link between the phantasmagoria and the mind became so commonplace that by midcentury physicians commonly used the operation of the magic lantern as a metaphor to explain the workings of the imagination.[70] Performances of delirium tremens gave a new and pressing meaning to phantasmagorical entertainment. By linking visions of the supernatural to a disease caused solely by heavy drinking, plays such as *The Drunkard* and *Ten Nights in a Barroom* created powerfully entertaining and highly symbolic performances.

Mesmeric performances also centered on images, even if the audience could not actually see them. In theatrical performances of mesmerism, the subject (almost always female) performed amazing mental feats that primarily involved "seeing" images. Under the direction of the "magnetizer" (almost always male), the subject read letters in sealed envelopes, described future events, and traveled in spirit to distant places, describing them in great detail.[71] Some Philadelphians had known of mesmerism since the 1780s.[72] Familiarity with the practice was common enough in scientific circles that the proprietor of the Philadelphia Museum, Charles Willson Peale, experimented with magnetism on sick family members in 1806.[73] Reports of the remarkable effects of mesmeric trance states always engendered deep skepticism, but discussion of the practice persisted in the medical community. At the Philadelphia Almshouse Hospital in 1839, for instance, physicians briefly experimented with mesmerism as a treatment for delirium tremens, reporting only limited success.[74]

In the 1840s alcoholic insanity, hallucinations, mesmeric trances, and other imaginative phenomena were irresistible topics of entertainment and speculation for middle-class Americans. At the heart of this fascination lay the compelling relation between the mental faculties of reason, imagination, and will. In his 1844 treatise on "mesmeric somniloquism," physician Daniel Drake drew on popular theatrical performances of mesmerism to speculate on the nature of the mind. Drake also had intimate knowledge of delirium tremens. One of the first physicians to publish an essay on the disease, he later vigorously warned of its horrors as a nationally recognized temperance advocate.[75]

The theory and practice of mesmerism changed over time, but in the 1840s mesmerists argued that humans live surrounded by an invisible magnetic fluid. Magnetizers projected their will to influence this fluid and place a willing and likely subject into a trance.[76] For Drake, the same mental operation involved in mesmeric phenomena also created the horrors of delirium tremens. Reflecting one common medical interpretation, Drake asserted that "all the phenomena" of mesmeric somniloquism "depend on the imagination." By submitting to the will of the magnetizer, he theorized, the mesmerized subject enters into a dream state, analogous to somnambulism. In a normal waking state, two factors keep the imagination under control. First, our external senses allow us to distinguish reality from images that well up from the imagination; and second, the faculties of reason and the will enable us to evaluate and act on the images presented to our mind. When in a somniloquent state, both external stimuli and the operations of reason and will become suspended, empowering the imagination. In theatrical mesmerism, "the whole mental performance is but the creation and exhibition of a series of phantasmagorical pictures" within the mind.[77]

Both the alcoholic maniac and the mesmerized subject lived in the imagination. Because "the sense of vision supplies the imagination with its principal materials," Drake wrote, "the magazines of the mind are almost filled with visual perceptions, and the imagination is not only the keeper of the key, but the artist which weaves them into tissues." This operation of the mind was the same that occurred in delirium tremens and ghost sightings more generally. Like the mesmerized subject, "the maniac from drink has an excited imagination, and enfeebled organs of sense, in which state he has many strange visions."[78]

Having abdicated reason and will, the mesmerized subject and the alcoholic maniac performed an identity ruled by imagination. As such, both figures lived beyond the psychic requirements of bourgeois selfhood, a position that invited prurient speculation. While delirium tremens signified depraved drinking habits, the lingering and titillating question raised by mesmerism was whether the subject would perform immoral acts in response to the magnetizer's suggestions. In the mesmeric drama, the performance gendered the imagination as female, supine under the power of the masculine will.

For their middle-class audiences, promoters tried to defuse the sexual connotations of the performances, for instance, with advertisements that showed the male magnetizer working over his subject while always under the interested but watchful gaze of a respectable middle-class family (fig. 14). Nevertheless, the appeal of the performances lay in their voyeuristic nature. The unscrupulous magnetizer who used his powers to render women sexually vulnerable persisted as a common character in antebellum popular fiction.[79] In the advertisement, the middle-class family watches fascinated by the power of the masculine will and the thrilling moral danger of the young woman, defenseless in her imaginative mental state, describing remarkable visions of far-off places, future events, and secret letters.

As demonstrated by *Ten Nights in a Barroom*, the middle-class voyeurism demonstrated in the advertisement shown in figure 14 also lay at the heart of the fascination with delirium tremens. The performance of the disease in the play flipped the gendered drama of mesmeric theater. Delirium tremens depicted the frightening power and danger of the masculine imagination, which could be properly constrained only by the combined powers of female domesticity and the Christian "heavenly Father."

Much of the drama in *Ten Nights in a Barroom* centers on the drunkard Joe Morgan. As a young man Joe had inherited a respectable business from his father, but he became impoverished through bad business decisions, a poor choice of friends, and a worsening drinking problem. In the play, Joe rots in a barroom, a loving father but "powerless, in the grasp of the demon."[80] One night Joe and the barkeeper, Simon Slade, get into an argument, and when the man throws a glass at Joe, he mistakenly hits Joe's innocent daughter Mary in the head, grievously wounding her. Lying in bed the next night, feverish, and drifting in and out of consciousness,

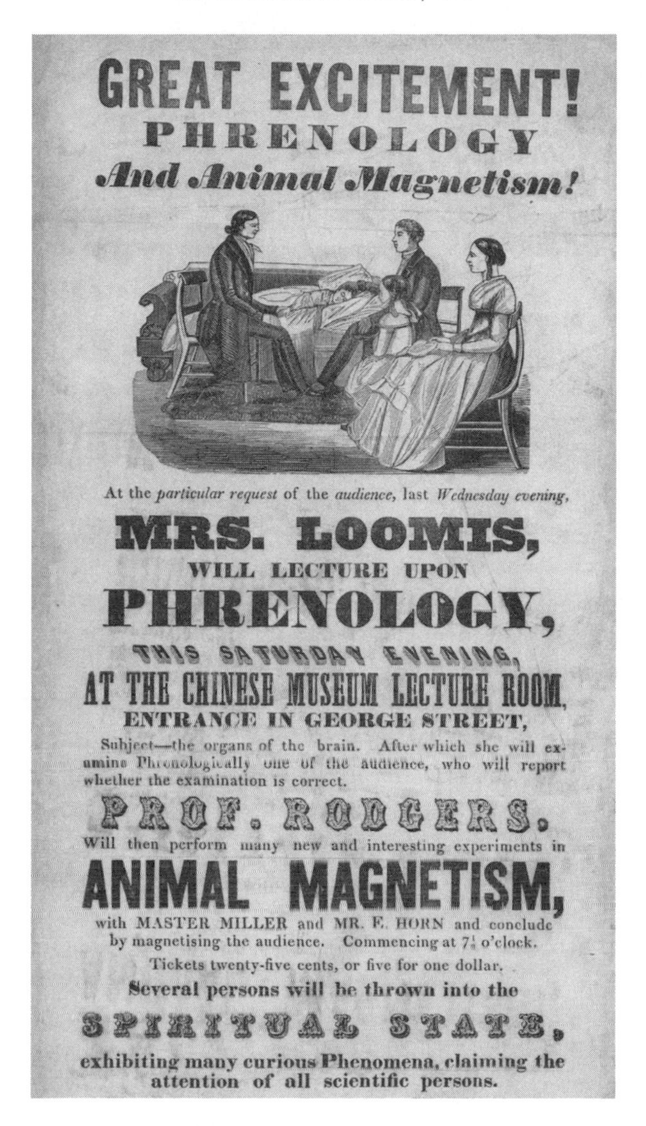

Figure 14. Advertisement for a mesmeric performance and lecture on phrenology in Philadelphia, 1843. Library Company of Philadelphia.

Mary begs her father not to go to the bar. Joe agrees and remains by his daughter's sickbed, but without his accustomed liquor he becomes more and more nervous.

Joe and Mary enter dream states simultaneously. Evoking images of a mesmeric performance, the girl envisions a happy future for her father.

Having signed the temperance pledge, he appears well dressed, happy, and healthful. The successful owner of a dry goods store, he embodies the middle-class ideals of the self-made man. She later recounts the vision to her father, saying, "I said, Oh! Father is this you? and then you took me up in your arms and kissed me, and said, Yes, Mary, this is your real father, not old Joe Morgan, but Mr. Morgan now."[81] At the same time that Mary has slipped into her trance, Joe becomes subject to the horrors of delirium tremens.

"Look—a huge snake is twining himself around my arms!" Joe yells. "How bright they look!—their eyes are glaring at me! And now they are leaping, dancing, and shouting with joy, to think the drunkard's hour has come. Keep them off! Keep them off! Oh! Horror! Horror!" As Joe descends into the drunkard's phantasmagoria, his daughter kneels in her sickbed and prays fervently, accompanied by soft music.[82]

In a magic-lantern slide depicting Joe suffering from delirium tremens (fig. 15), the horrified man stares into the maelstrom. Like the somnambulist or the mesmerized subject, his imagination overpowers his will and reason, leaving him at the mercy of inner demons. In some magic-lantern adaptations another slide followed this image of Joe's hallucination. That slide has not survived, but one playbill described it as a vision of Satan sitting on his throne, perhaps similar to a slide titled "Delirium Tremens" (fig. 16), painted by a Philadelphia artist about the turn of the century. The slide depicts the snakes and eyes in Joe's ravings, suggesting the artist was inspired by the popular play.[83]

Both Joe's and Mary's dreams draw heavily on mesmeric symbology to moralize about the dangers of the unrestrained imagination. Evoking the relationship between male magnetizer and female subject, "Our Heavenly Father" guides the daughter's imagination to envision the promise of temperance, while Joe's alcoholic imagination overpowers his own degraded will, throwing him into hell. Mrs. Morgan relegates Joe's waking nightmares to their appropriate domain when she gives him a dose of opium, putting him to sleep. Opium was the home remedy for delirium tremens in midcentury middle-class ladies' housekeeping and etiquette guides.[84] Waking in the morning, his imagination safely under his wife's dominion, Joe promises his daughter he will never drink again. Her angelic work on earth completed, the daughter announces that God has called her back to heaven and dies.

In part, *Ten Nights in a Barroom* demonstrates how delirium tremens, and more generally medical theories and practices surrounding alcohol abuse, functioned as a nightmare, rationalizing why some men failed to

Figure 15. Magic-lantern slide depicting Joe Morgan suffering delirium tremens. Joseph B. Beale, manufactured ca. 1880 by the T. H. McAllister Company, New York. Image courtesy of the American Magic-Lantern Theater.

attain the middle-class dream. While writers like Lippard used the association of delirium tremens with moral and social downfall to criticize the depravity inherent in the antebellum class structure, Arthur depicted Joe's disease as affirming America's middle-class promise. *Ten Nights in a Barroom* does not cast Joe's business failure and the poverty of his previously respectable family as resulting from the common problems associated with the antebellum market economy, including economic depressions, unstable banks, or rapacious capitalists. Rather, Joe's failure results from an individual, psychological struggle with his inner alcoholic demons.

In Poe's "Maelström," the fisherman's "old mates and daily companions" pull him on board, but Joe's refuge is the middle-class cult of domesticity, not a working-class brotherhood of former drunkards. Salvation comes to Joe when he abandons the masculine world of the tavern for the protection of the feminine domestic sphere. This move is illustrated in the

Figure 16. "Delirium Tremens." Magic-lantern slide depicting hallucinations typical of the disease. Joseph B. Beale, ca. 1900. Image courtesy of the American Magic-Lantern Theater.

magic-lantern slide (fig. 15) by Mary's tiny hand gently pulling on his elbow, urging him to turn toward the light. Mr. Morgan goes on to fulfill his daughter's prophecy, regaining his divinely sanctioned respectability. In the final act of the play, Mr. Morgan sits with his wife in an elegant parlor. Reflecting on their success, Mrs. Morgan remarks that the ten years since Mary's death "have rolled by like some sweet dream."[85] Mr. Morgan's cure reaffirms that economic prosperity, middle-class respectability, and domestic bliss were available to any man of temperate and industrious habits.

Ten Nights in a Barroom illustrates the psychic conflicts at the heart of the American middle-class ethos at midcentury. The attraction of the horrors of delirium tremens derived from a repressed and subversive longing for something beyond the dry rigor of polite society and the terrifying dangers of the capitalist market. Tempting the mind's dark nature became an alluring process of self-cultivation in which a man confronted horrors awakened by alcoholic compulsion. Joe hallucinates serpents, a Christian symbol of the temptation for forbidden knowledge. As with Poe's old fisherman, the play suggests that to be subject to the hideous imaginings of delirium tremens is to flirt with the abyss and glimpse its marvels. As was typical of Washingtonian-influenced narratives, Mr. Morgan's experiences enable him to speak with great moral authority as the play closes with an appeal for temperance reform. Reflecting on his descent into depravity, Mr. Morgan sounds like the fisherman when he says,

> I have seen frightful death-bed scenes, where the frothing lip and the blood-shot eye, the distorted features and the delirious shrieks, told the fierce agony of the departed soul, and as my shuddering glance takes in but a feeble

Figure 17. Magic-lantern slide depicting Mr. Morgan addressing the townspeople. Joseph B. Beale, manufactured ca. 1880 by the T. H. McAllister Company, New York. Image courtesy of the American Magic-Lantern Theater.

> outline of the revolting spectacle, I know how much of the great sea of human crime, and want, and woe, pour through the slender channel of that one word—Drunkard.[86]

Joe Morgan's nightmare lets him step outside the middle-class parlor to witness a hideously fascinating underworld. Arthur contains the subversive implications of delirium tremens's allure by casting Mr. Morgan as a newly authoritative voice for moral reform (fig. 17). He rallies the townspeople to reject the evils of drink and drive out Simon Slade, freeing the town from the social, economic, and political ills brought on by alcohol. Joe Morgan's suffering becomes a beacon warning the community away from the temptations of liquor.

Gothic horror and romantic longing thus characterized middle-class fascination with pathological drinking. In midcentury popular culture, portrayals of delirium tremens contained the same sense of desire that shaped physicians' earliest narratives. The stunning popularity of *Ten Nights in a Barroom* and *The Drunkard* speaks to the depths of these dark currents in mass culture. Morgan attains the middle-class "dream" through his daughter's angelic vision and his wife's patient domestic skill, but his shamanic journey through the nightmare makes him a man of greater psychological fortitude, a man who has confronted, fought, and mastered his inner demons. Theatrical performances of delirium tremens appealed to these repressed desires to escape the parlor, wallow in dark shadows, and experience something of eternity.

EPILOGUE

Alcoholics and Pink Elephants

At the turning point in Walt Disney's film *Dumbo,* a lonely young circus elephant finds some relief from his despair in alcohol. With no father and unjustly separated from his mother, Dumbo has been forced to fend for himself. But he is too small and too awkward to perform with the stronger, more graceful elephants, mostly because of his absurdly large ears. Frustrated, the ringmaster puts him in the clown act. This embarrassing demotion shocks the other elephants, who shun him to protect their own respectability. Dumbo finds himself alone in the world save for one unlikely ally: a streetwise mouse named Timothy. Outraged at Dumbo's treatment, Timothy resolves to teach the naive youth the ins and outs of circus life. One night, after visiting his imprisoned mother, Dumbo is inconsolable and develops hiccups from crying. Timothy offers him a drink from a bucket, not noticing that the water bubbles ominously, spiked with champagne by the raucous clowns. Dumbo and Timothy both become pleasantly intoxicated.

In one of the most memorable sequences in all of Walt Disney's films, Dumbo and Timothy embark on an alcoholic fantasy as Technicolor elephants march across their vision. The hallucinations build until they are engulfed in swirling, phantasmagorical figures and a lush musical score. Initially the images are frightening, accompanied by haunting laughter and lyrics:

> Look out! look out!
> Pink elephants on parade
> Here they come!
> Hippety hoppety
> They're here and there
> Pink elephants everywhere

But soon the scenes and music turn playful and exciting, with lightning bolts, mambo rhythms, and graceful ice dancing. What began as a nightmare turns into an exotic dream world that transports the two far beyond the muddy, mundane circus compound into a world of fantastic entertainment.[1]

Released in 1941, *Dumbo* established pink elephants as the standard cliché for delirium tremens, but in the twentieth century the disease no longer dominated representations of pathological drinking as it once had. Although the affliction remained a common diagnosis in American hospitals, popular concern focused on the more general disease called alcoholism. In the 1930s and 1940s, physicians, public health officials, and social reformers developed and disseminated new theories of alcohol addiction with the influential support of a new mutual aid society called Alcoholics Anonymous. Unlike earlier temperance advocates, the group did not dwell on graphic descriptions of the horrors of delirium tremens, perhaps in part because its long-standing presence in popular culture had robbed the disease of its power to frighten. That allusions to delirium tremens began to appear in children's cartoons illustrates how distant dramatizations of the disease had become from its ugly reality.

Dumbo and Timothy were not the first drinkers to see phantasmagorical pachyderms. In the autobiographical *John Barleycorn, or Alcoholic Memoirs* (1913), Jack London wrote that common gutter drunks saw "blue mice and pink elephants." By contrast, drinkers with "imagination and vision" succumbed to intellectual "spectres and phantoms that are cosmic and logical." Viewing these "spectral syllogisms," the imaginative drinker contemplates the "iron collar of necessity" that alcohol has "welded about the neck of his soul." For London, alcoholism was death by "suicide, quick or slow, a sudden spill or a gradual oozing away through the years" that no alcoholic ever escapes.[2] The first flying elephant was more playful than London's, though on black-and-white film its color is ambiguous. In the Prohibition-era cartoon "Felix Finds Out" (1924), Felix the Cat sets out to discover what makes the "moon shine." His travails lead him to sample a jug of moonshine, and he is then haunted by a Chinese dragon and a winged elephant.[3]

Creeping into children's cartoons, alcoholic hallucinations carried many of their nineteenth-century meanings. Following the form of popular temperance dramas, perhaps most famously *Ten Nights in a Barroom*, the confrontation with alcoholic phantasms precipitated a change in Dumbo's life. Dumbo and Timothy wake up in the upper branches of a very tall tree, and a flock of crows is laughing at them. With no memory of what happened and perplexed about how they got up so high, Timothy reasons that Dumbo must have taken flight using his huge ears. With the help of the jazz-singing crows, who are thinly veiled minstrel show characters, Dumbo transforms his embarrassing ears into wings. He triumphantly returns to the circus as the world's first flying elephant, takes revenge on his tormentors, rescues his mother, and attains instant celebrity. In the end he flies into a sunny future while his mother sits happily in their shiny, streamlined railcar.[4] Disney animated pink elephants to dramatize a story of personal transformation appropriate for small children, only hinting at the terror delirium tremens had evoked in the previous century.

Themes and tropes of an earlier era thus found new resonance. Even if delirium tremens had faded in the cultural imagination, the disease continued to inform how Americans conceived of pathological drinking. As illustrated in Dumbo's early failures as a circus elephant, allusions to delirium tremens continued to center on white men's struggle for economic success. But even so, these stories were no longer constrained by temperance ideology. Early temperance societies had stressed that thrift, industry, Christian piety, hard work, and healthy habits were all crucial to attaining wealth. *Dumbo* dispensed with these austere and banal requirements, turning instead to inner self-discovery and revelation. An accidental descent into alcoholic insanity reveals Dumbo's hidden talent, bringing fame and fortune. Success, Disney teaches us, just takes a little self-confidence. Dumbo's rise from failed circus elephant to national celebrity is the just outcome for the pain he suffered during his agonizing separation from his mother and his humiliating performances with clowns. That black jazz musicians help him toward this revelation further highlights Dumbo's need to seek his true self outside the competitive arena of career and work and the stultifying confines of respectable elephant society. The perverse fascination that characterized mid-nineteenth-century attitudes toward pathological drinking still lingered. In American mass culture, delirium tremens remained compelling theater because it continued to express a romantic longing to break free of the painful strictures of middle-class existence. Still a deadly affliction, delirium tremens came to offer a cure.

Toward Alcoholism

The theories of alcoholism advanced in the 1930s and 1940s were the culmination of decades of work by addiction physicians, a medical specialty that first took shape after the Civil War. These new medical responses to alcohol abuse continued to reflect the broader cultural investment in the belief that men attained wealth and respectability through a rational pursuit of economic gain, proper moral conduct, and masculine will. "Even more than they had in antebellum America," writes historian Scott Sandage, "the self-made man and the broken man represented the poles of an ideology of manhood based on achieved identity—the conviction that all men earned their fates and thus deserved whatever credit or disgrace they accrued."[5] In the 1870s, an illustration from a popular phrenological journal demonstrated the long resonance of antebellum physicians' campaign to tie the sciences of phrenology and physiology to this ethos (fig. 18). Reprinted in the *Pennsylvania School Journal*, a publication of the Pennsylvania State Education Association, "Two Pictures" presents the physiognomies of temperance and intemperance from cradle to grave. Each picture shows progressive changes in the countenance of a man, one temperate and one intemperate. The life of the abstemious man climaxes in a noble, wise, and friendly face with a lush beard. The intemperate man grows bloated and diseased—ultimately dead in a gutter. Similar popular illustrations dated back at least to the eighteenth century and William Hogarth's series of engravings "A Rake's Progress." "Two Pictures" drew on this tradition to graphically illustrate the popular health literature of the day. Titling the two halves of the illustration "A Normal Life" and "An Abnormal Life," the commentary declared, "We cannot ignore the laws which govern human life and escape the consequences." Physicians transmuted self-made manhood into a science of human life.

But in the postbellum era the explanatory power of phrenology and physiology receded. Intellectual developments and professional pressures drove changing medical conceptions of pathological drinking. As the profession became more specialized, some doctors capitalized on the broad public concern with the social consequences of heavy drinking to shape new professional opportunities. Most often with public funding, physicians, religious activists, and social reformers built large asylums for the treatment of inebriates. In keeping with other Progressive Era reform movements, asylum advocates sought to apply scientific principles and professional expertise to a pressing social problem. Before the Civil War, the great majority of physicians had aspired to private practices catering to respectable clients, even if that ideal was difficult to attain. In the late nineteenth century, some doctors found that working in large institutions brought a more stable career path and a measure of professional recognition and prestige. Inebriate asylums became laboratories as well as treatment facilities, as doctors applied new theories of the body and mind to the problem of alcohol addiction. Variously termed monomania, dipsomania,

Figure 18. "Two Pictures," *Pennsylvania School Journal* 24, no. 8 (February 1876): 286. Periodical Collection, Gutman Library, Harvard Graduate School of Education.

inebriety, and alcoholism, disease models for compulsive drinking grew out of intellectual trends such as hereditarian racial science and, by the early twentieth century, Freudian psychology.[6] The study of alcohol addiction was slow to eclipse the long-standing medical preoccupation with delirium tremens, however. In 1880, for instance, the accumulated medical literature on delirium tremens still dwarfed the emerging progressive literature on inebriety.[7]

Historians still debate whether Prohibition improved pubic health.[8] Its repeal in 1933, however, discredited the moral reformers who had made alcohol a compelling political issue. Efforts to quell the political passions surrounding alcohol empowered addiction professionals.[9] New behavioral norms had developed in the dimly lit speakeasies of the 1920s, and in the 1930s and 1940s drinking alcohol, even liquor, became far more socially acceptable for both men and women. The experience of Prohibition convinced many that eliminating alcohol from society was impossible and unnecessary, given that most people could drink moderately. The worst consequences of drinking seemed to manifest themselves in only a small minority of people. Eager to identify all the negative social consequences of alcohol abuse with this minority, the alcohol industry invested in research into pathological drinking. The Yale Center for Alcohol Studies, directed by E. Morton Jellinek, was especially influential in applying Freudian psychology in a new disease model. "Alcoholism" defined problem drinkers as suffering from a psychological disorder or deficiency that made them unable to drink moderately.[10]

This new disease model came to dominate academic research, social responses, and cultural representations of pathological drinking.[11] "Alcoholism" still persists in common parlance, even if addiction researchers have long since discredited the original disease model developed at the Yale Center. This persistence can be attributed in large part to the success of Alcoholics Anonymous. Founded in 1935, AA grew to be an international mutual aid society made up of tens of thousands of loosely affiliated meetings and millions of adherents. AA's twelve-step recovery program has become the most common template for efforts to treat alcoholics, and it has been adapted to treat other modern and ancient compulsions such as pathological gambling, sex, eating, and Internet gaming. Not strictly scientific, AA describes alcoholism as a disease, but one that has physical, mental, and spiritual dimensions. Throughout the twentieth century, AA has cooperated with medical researchers toward their common goal of helping alcoholics. Through AA, researchers have developed new paradigms for understanding the underlying pathology of alcoholism and have worked to publicize recent theoretical developments in addiction science.[12]

As public and medical concern with problem drinking focused heavily on the compulsion to drink, delirium tremens came to be seen as one extreme consequence of this more fundamental medical condition.[13] Delirium tremens remained an object of medical research in the twentieth century, albeit a relatively minor one. Perhaps the most significant and remarkable experiment was one that aimed to settle the century-long debate over whether delirium tremens was caused by withdrawal or just by heavy drinking. In 1953 researchers conducted the experiment on ten male prisoners. Morphine addicts serving time in a penitentiary for possession, they had all abstained from alcohol and drugs for at least three months. They volunteered for the experiment, in which they were kept heavily intoxicated for periods ranging from seven days to almost three months. Each received average daily doses of from nine ounces to over sixteen ounces of 95 percent ethyl alcohol—the equivalent of 190-proof liquor. Physicians were amazed at their subjects' tolerance: the daily alcohol intake of four of the ten men approached what doctors had believed was the absolute maximum any human could endure.

Four subjects developed severe withdrawal symptoms. As in nineteenth-century case histories, physicians carefully recorded their hallucinations, even though the authors did not discuss the visions as medically relevant. These hallucinations were not entirely different from a Disney cartoon, though generally more disturbing. Dumbo saw disembodied, devilish elephant heads emitting spooky laughter, for instance. After eighty-seven days of continual intoxication, "Al" saw, among other terrifying things, a "disembodied head which was shrunken and had the appearance of heads prepared by a tribe of South American Indians. The eyes of this head followed the patient as he moved in bed. On closing his eyes, he saw a dwarf who would disappear whenever he opened his eyes." Intoxicated for seventy-eight days, "Jack" was haunted by voices immediately after he stopped. His hallucinations worsened on the third day. Just as Dumbo discovered he had the gift of flight, Jack described hallucinations of "his bed flying through the air, going through dark tunnels, and so forth." Jack's dream lacked the joy and triumph of the cartoon, as "he described being attacked by an imaginary animal which spat acid in his face. He would strike at the animal with his pillow and said that he had caught it several times." The study concluded that sudden abstinence was at least one factor that precipitated delirium, tremors, convulsions, and other symptoms.

After the experiment ended, several of these human subjects required as much as six weeks of medical care before they were well enough to return to their prison cells.[14]

Technicolor Pachyderms

In American popular culture, representations of delirium tremens became ever more distant from the terrifying and life-threatening reality experienced by Jack and Al. Today, for instance, Delirium Tremens is the name of a popular Belgian beer, which features a pink elephant on the label. That a company would choose to market a beer by that name indicates how far the disease has come from its first description in the *Edinburgh Medical and Surgical Journal*. Delirium tremens nevertheless retained many of the symbolic meanings that first became associated with alcoholic insanity in the 1810s and 1820s.

Cinematic dramatizations of delirium tremens appeared in sentimental portrayals of alcoholism such as D. W. Griffith's *The Struggle* (1931) and Billy Wilder's *The Lost Weekend* (1945). Typically, delirium tremens struck at the nadir of the alcoholic's suffering. In the movie *Days of Wine and Roses* (1962), for instance, young Joe Clay is driven to alcoholism by the pressures of his job as a high-profile public relations man. Drinking destroys his career and family. He winds up incarcerated in a medical facility, writhing in a violent delirium, restrained by a straitjacket. This alcoholic nightmare marks the turning point in Joe's life. Shortly after he recovers, he joins Alcoholics Anonymous and begins a road to recovery. The gender politics of the mid-twentieth century are far different from those of the mid-nineteenth century.[15] In stark contrast to the typical Washingtonian narrative, the tragedy in *Days of Wine and Roses* centers on Joe's wife Kirsten. She has also fallen prey to the bottle, a habit Joe introduced her to. While Joe's recovery enables him to care for his child, his wife cannot face the social stigma of declaring herself an alcoholic and so cannot follow him into sobriety.[16] The movie nevertheless preserves the nineteenth-century construction of delirium tremens as primarily a masculine drama. Kirsten's alcoholism develops in the home, expressing her inability to reconcile the conflicting demands of being a devoted daughter, a good mother, and the faithful wife of an alcoholic. She never develops delirium tremens. While Joe writhes in his padded cell under the care of physicians and hospital staff, Kirsten is drunk in her father's house. He forces her into the shower to sober her up.

Even the sanitized realism of the hospital scenes in *Days of Wine and*

Roses is rare. Popular culture purged delirium tremens of the violent vomiting, diarrhea, seizures, and other ugly symptoms known to hospital workers. Most often, fanciful spectral visions and dreams allude to the pathological condition. Nevertheless, the disease retains some of its association with terror and radical evil. In Stanley Kubrick's film adaptation of Stephen King's *The Shining* (1980), for instance, a drunkard father, Jack Torrance, is beset by ghosts while wintering with his family in an isolated, empty hotel. While Jack sinks into insanity, agonizing over his stuttering career and struggling with his craving for liquor, his imaginative son suffers frightening visions of the past and an impending bloody terror. The boy's clairvoyance eventually enables him to escape with his mother. His fixation on "red rum," a mysterious phrase from his dreams that foretells his father's intentions, and Jack's conversation with a spectral bartender strengthen the association of alcoholic depravity, delirium tremens, and pathological murder.[17] *The Shining* draws on delirium tremens in the gothic lexicon popularized by Edgar Allan Poe, George Lippard, T. S. Arthur, and others. The murderous Jack Torrance evokes such compelling nineteenth-century villains as Pap, the drunk and ne'er-do-well father of Mark Twain's Huckleberry Finn, who tries to murder Huck while suffering wild hallucinations.

Other representations of delirium tremens lack the horror evoked when John Gough first trembled and thundered on the antebellum stage. In the twenty-first-century movie adaptation of the *Adventures of Tintin* (2011), for instance, the drunkard Captain Haddock becomes desperate for liquor after he and Tintin are marooned in a desert, a landscape that serves as a metaphor for temperance. Growing distraught without his liquor, the captain suddenly sinks into an elaborate mirage: a violent sea battle, pitting his ancient ancestor Captain Haddock against the villainous pirate Red Rackham. This vision in the desert enables the captain, previously a drunkard at the mercy of an unscrupulous villain, to unlock the mystery of the model ship *Unicorn* and so reclaim his honor and his family's fabulous wealth.

As enacted by Dumbo, Captain Haddock, Jack Torrance, and Joe Clay, delirium tremens remains a performance dramatizing the psychic dimension of the male drunkard's struggle with depraved and destructive impulses. In these cinematic representations, men suffer disappointment and despair because of their failures in the marketplace. Their alcoholic visions are transformative, however, allowing the protagonists to come out of themselves and transcend their limitations. Casting economic failure as psychological struggle, these stories deny the social context of drinking and despair. The struggles of these characters are not the consequence of social inequality, exploitation, or economic turmoil. Drink is the architect of Joe

Clay's demise, not the inherent instability and chaos of the competitive market. Further, none of these twentieth-century narratives leads to the birth of a new John Gough. After his illness, Joe Clay does not feel morally compelled to agitate against the mid-twentieth-century liquor industry. Likewise, Dumbo does not deliver a temperance lecture to the bibulous circus clowns. Dumbo's success is his achieving celebrity, a dream he attains through knowledge gained in an alcoholic nightmare. The disease no longer carries broader social implications. In the nineteenth century, authors such as George Lippard used the disease as a metaphor to condemn the depravity of the wealthy and the immorality inherent in the shocking disparities between rich and poor. With the failure of Prohibition, delirium tremens lost this symbolic dimension.

Journeying to the Mountaintop

In the mid-twentieth century, delirium tremens became a window into an entirely individual transformation, an avenue to intense experience that spoke to a larger element in popular culture: a longing for psychic rebirth.[18] In the second half of the twentieth century, Americans gained many more doorways into the land of Technicolor pachyderms and phantasmagorical dreamscapes as their recreational pharmacopoeia expanded dramatically. With its swirling, colorful apparitions, the pink elephant sequence in *Dumbo* evokes something of the psychedelic culture of the 1960s. Consider that apostles of the psychedelic revolution, such as Ken Kesey's Merry Pranksters, could easily have written the lyrics:

> I can stand the sight of worms
> And look at microscopic germs
> But Technicolor pachyderms
> Is really too much for me.
>
> I am not the type to faint
> When things are odd or things are quaint
> But seeing things you know that ain't
> Can definitely give you an awful fright.
>
> What a sight!
> Chase 'em away! Chase 'em away!
> I'm afraid, need your aid
> Pink elephants on parade.

Two years before the discovery of the psychedelic properties of LSD in 1943, Disney used alcoholic hallucinations to celebrate the transformative potential of "seeing things you know that ain't."[19]

Psychiatrists made this comparison in the earliest years of LSD's existence. Some of the first practical experiments with LSD, in fact, used the drug on alcoholics in an attempt to replicate the psychic experience of delirium tremens. The idea was not just to scare them sober but to artificially induce a profound awakening. In 1953 psychiatrists Humphry Osmond and Abram Hoffer had been conducting research into the effects of LSD but were unclear on the drug's significance or its potential application. It occurred to them that the patient narratives they had gathered in their LSD experiments bore a strong resemblance to the stories alcoholics told about delirium tremens. Contemporary addiction research also noted that while delirium tremens was life threatening, the experience of the disease often led the drinker to finally choose to stop drinking. In their subsequent experiments, Hoffer and Osmond found that LSD caused alcoholic patients to have a similarly profound experience, a spiritual or transformative awakening. Historian Erika Dyck writes that "alcoholic patients responded extraordinarily well to the LSD treatments, convincing Osmond and Hoffer that the psychedelic experience itself conveyed potential therapeutic benefits."[20] The thrust of Osmond and Hoffer's application of LSD to alcoholism, however, ran counter to much of the science of addiction. As with Dumbo's pink elephant dream, Dyck writes that patients' "experiences of personal insight and reflection often defied scientific explanation."[21]

Patients described their LSD experiences as positive and therapeutic. As Jack Kerouac explained, the horrors of delirium tremens are "not so much a physical pain but a mental anguish so intense that you feel you have betrayed your very birth, the efforts nay the birth pangs of your mother when she bore you and delivered you to the world." Even so, for Kerouac delirium tremens did bear some resemblance to a psychedelic experience. His experiments with drugs had not prepared him, however.

> No matter how many books on existentialism or pessimism you read, or how many jugs of vision-producing Ayahuasca you drink, or Mescaline take, or Peyote goop up with—that feeling when you wake up with the delirium tremens with the *fear* of eerie death dripping from your ears like those special heavy cobwebs spiders weave in the hot countries, the feeling of being a bentback mudman monster groaning underground in hot steaming mud pulling a long hot burden nowhere, the feeling of standing in hot boiled pork blood . . .

The commonality between the psychedelic experience and delirium tremens is the perspective gained through transcending the self. Kerouac continues in his stream-of-consciousness mode, "The face of yourself you see in the mirror with its expression of unbearable anguish so hagged [*sic*] and awful with sorrow you cant [*sic*] cry for a thing so ugly, so lost, no connection whatever with early perfection and therefore nothing to connect with tears or anything."[22] This altered and extreme state of mind offers an opportunity for self-reflection and self-knowledge, and perhaps the possibility of a new way of being.

A new way of being is the central goal of the AA twelve-step program. Intrigued by the research of Osmond and Hoffer, AA's founder Bill Wilson came to believe that LSD had the potential to help alcoholics achieve this goal. Wilson first took the drug in 1956, and he described the experience as closely resembling his own spiritual awakening, which had enabled him to achieve sobriety after years of struggle. The story of Wilson's awakening remains one of the central texts of AA and thus one of the most significant recovery narratives published in the twentieth century. In the winter of 1934, Wilson had checked in to a New York hospital for inebriates and drug addicts. "Treatment seemed wise, for I showed signs of delirium tremens," he wrote.[23] While being treated for his symptoms with a tincture of belladonna, a drug that can produce vivid hallucinations, Wilson had a profound spiritual experience. In his telling, it was triggered at the moment he was able to surrender his will to a higher power: a God of his own conception. "It meant destruction of self-centeredness," he wrote. The moment he had fully come to terms with his new program, "there was a sense of victory, followed by such a peace and serenity as I had never known. There was utter confidence. I felt lifted up, as though the great clean wind of a mountain top blew through and through."[24] Wilson's mountaintop experience became enshrined in the AA program, which holds that a spiritual awakening is crucial to recovery. Believing that LSD could help others have a similar experience, Wilson experimented with the drug into the 1960s. Although he initially believed all alcoholics should have access to the drug, he never publicly promoted the treatment.[25]

Historians of AA have written much about Wilson's spiritual awakening and its evocation of various strains of Protestant piety and popular mysticism.[26] My purpose here is not to question the validity of the story or deny the significance of religion in shaping this experience. Rather, I want to highlight the echoes of imagery and themes that took shape around representations of delirium tremens. Wilson's story evokes many comparisons to nineteenth-century temperance narratives. He was an ambitious young businessman who, in the treacherous economic climate of the 1920s, fell prey to failure and drink. His struggle to overcome alcoholism occurred during a profound economic depression. The cure for his affliction was a new selfhood, an entirely new way of being, which he finally achieved through a transformative dream while suffering the symptoms of delirium tremens. One reason Bill Wilson's story has been meaningful to so many is that the plot has such deep roots in American popular culture.

Pathological drinking has always had this romantic drama at its heart. It was evident in the earliest stories told about delirium tremens in Philadelphia's anatomy theaters and lecture halls, amid the economic wreckage of the Panic of 1819. Then, ambitious young men, anxious about their own social position and economic future and subjecting themselves to the self-discipline required by the competitive marketplace, marveled at the diseased imaginings of bankrupts and losers. By the twentieth century, delirium tremens continued to speak to a yearning to travel into the mysteries of the psyche in search of self-knowledge, authentic experience, and social success. Dumbo's alcoholic dream enables him to literally fly to new heights and discover his true self. For Wilson and others this romantic quest had become more than a cultural trope popularized on the antebellum stage; it had become a way some people suffering alcohol addiction could make sense of their struggle with the affliction. Narrating the journey beyond the self to wrestle with inner psychic demons enabled some to find a new path forward.

NOTES

INTRODUCTION

1. "Poe's Last Visit to Philadelphia," Society Collection, Poe Letters, Historical Society of Pennsylvania, Philadelphia; Kenneth Silverman, *Edgar A. Poe: Mournful and Never-Ending Remembrance* (New York: Harper Perennial, 1991), 415–19.
2. Silverman, *Edgar A. Poe*, 415–16.
3. "Poe's Last Visit to Philadelphia," 2.
4. "Fugitive Poetry of America," *Southern Quarterly Review* 14, no. 27 (1848): 119.
5. Isaac Clarkson Snowden, "On Mania á Potu," *Eclectic Repertory and Analytical Review* 5 (1815): 372–92.
6. Silverman, *Edgar A. Poe*, 433–37.
7. David C. Dugdale, MD, "Delirium Tremens," US National Library of Medicine and the National Institutes of Health, http://www.nlm.nih.gov/medlineplus/ency/article/000766.htm.
8. David B. Merill, MD, "Alcohol Withdrawal," US National Library of Medicine and the National Institutes of Health, http://www.nlm.nih.gov/medlineplus/ency/article/000764.htm.
9. Harris Isbell, H. F. Fraser, Abraham Wikler, R. E. Belleville, and Anna J. Eisenman, "An Experimental Study of the Etiology of 'Rum Fits' and Delirium Tremens," *Quarterly Journal of Studies on Alcohol* 16, no. 1 (1955): 1–33.
10. Ric N. Caric, "Hideous Monsters before the Eye: Delirium Tremens and Manhood in Antebellum Philadelphia," in *Gender, Health, and Popular Culture*, ed. Cheryl Krasnick Warsh (Waterloo, ON: Wilfred Laurier University Press, 2011); Ric N. Caric, "The Man with the Poker Enters the Room: Delirium Tremens and Popular Culture in Philadelphia, 1828–1850," *Pennsylvania Magazine* 74, no. 4 (2007): 452–91.
11. John Romano, "Early Contributions to the Study of Delirium Tremens," *Annals of Medical History*, ser. 3, no. 3 (1941): 128–39.
12. Editor's introduction in Thomas Trotter, *An Essay Medical, Philosophical, and Chemical on Drunkenness, and Its Effects on the Human Body*, ed. Roy Porter, Tavistock Classics in the History of Psychiatry (New York: Routledge, 1988), xii.
13. The literature on the relation between disease and history is vast. Among many works, several recent essays that have influenced this book include Ludmilla Jordanova, "The Social Construction of Medical Knowledge," in *Locating Medical History: The Stories and Their Meanings*, ed. Frank Huisman and John Harley Warner

(Baltimore: Johns Hopkins University Press, 2007), 346; Charles E. Rosenberg, "What Is Disease? In Memory of Owsei Temkin," *Bulletin of the History of Medicine* 77, no. 3 (2003): 491–505; Mary E. Fissell, "Making Meaning from the Margins: The New Cultural History of Medicine," in *Locating Medical History: The Stories and Their Meanings*, ed. Frank Huisman and John Harley Warner (Baltimore: Johns Hopkins University Press, 2007).

14. Joseph Gusfield, *Symbolic Crusade: Status Politics and the American Temperance Movement*, 2nd ed. (Chicago: University of Illinois Press, 1986); Ian Tyrrell, *Sobering Up: From Temperance to Prohibition in Antebellum America* (Westport, CT: Greenwood Press, 1979); Ronald G. Walters, *American Reformers, 1815–1860* (New York: Hill and Wang, 1978), 123–43.
15. Tyrrell, *Sobering Up*; Mary P. Ryan, *Cradle of the Middle Class: The Family in Oneida County, New York, 1790–1865* (New York: Cambridge University Press, 1981), 105–44; Stuart Blumin, *The Emergence of the Middle Class: Social Experience in the American City, 1760–1900* (New York: Cambridge University Press, 1989), 195–205; Paul Johnson, *A Shopkeeper's Millennium: Society and Revivals in Rochester, 1815–1837* (New York: Hill and Wang, 1978); Robert Abzug, *Cosmos Crumbling: American Reform and the Religious Imagination* (New York: Oxford University Press, 1994), 81–104; Herbert Gutman, *Work, Culture, and Society in Industrializing America* (New York: Knopf, 1976).
16. Elaine Franz Parsons, *Manhood Lost: Fallen Drunkards and Redeeming Women in the Nineteenth-Century United States* (Baltimore: Johns Hopkins University Press, 2003), 9. Less dismissive but still skeptical is Bruce Dorsey, *Reforming Men and Women: Gender in the Antebellum City* (Ithaca, NY: Cornell University Press, 2002), 116–17.
17. In one of the few studies to look at these developments, Stephen Nissenbaum's *Sex, Diet, and Debility in Jacksonian America: Sylvester Graham and Health Reform* (Chicago: Dorsey, 1980) sketches how intellectual developments within the Philadelphia medical community shaped the thinking of temperance activist and health reformer Sylvester Graham.
18. Susanna Barrows and Robin Room, "Introduction," in *Drinking: Behavior and Belief in Modern History*, ed. Susanna Barrows and Robin Room (Berkeley: University of California Press, 1991), 12.
19. W. J. Rorabaugh, *The Alcoholic Republic: An American Tradition* (New York: Oxford University Press, 1979), 38.
20. Abzug, *Cosmos Crumbling*, 94–95. Also see Scott C. Martin, *Devil of the Domestic Sphere: Temperance, Gender, and Middle-Class Ideology, 1800–1860* (DeKalb: Northern Illinois University Press, 2008), 68–86.
21. John Bynum, "Alcoholism in the First Half of the Nineteenth Century" (PhD diss., Yale, 1969); Romano, "Early Contributions to the Study of Delirium Tremens," 128–39; Katherine H. Nelson, "The Temperance Physicians: Developing Concepts of Addiction" (PhD diss., American University, 2006).
22. Charles E. Rosenberg, *The Care of Strangers: The Rise of America's Hospital System* (New York: Basic Books, 1987), 63; William Barlow and David O. Powell, "A Dedicated Medical Student: Solomon Mordecai, 1819–1822," *Journal of the Early Republic* 7, no. 4 (1987): 377–97; James Webster, *Facts concerning Anatomical Instruction in Philadelphia* (Philadelphia: James Webster, 1832); Michael Sappol, *A Traffic of Dead Bodies: Anatomy and Embodied Social Identity in Nineteenth-Century America* (Princeton, NJ: Princeton University Press, 2002), 112–17; Simon Baatz, "'A Very Diffused

Disposition': Dissecting Schools in Philadelphia, 1823–1825," *Pennsylvania Magazine of History and Biography* 108, no. 2 (1984): 206.

23. Rorabaugh, *Alcoholic Republic*, 38; Harry G. Levine, "The Discovery of Addiction: Changing Conceptions of Habitual Drunkenness in America," *Journal of Studies on Alcohol* 39, no. 1 (1978): 143–74; *The Anniversary Report of the Managers of the Pennsylvania Society for Discouraging the Use of Ardent Spirits* (Philadelphia: Pennsylvania Society for Discouraging the Use of Ardent Spirits, 1831), 68.
24. For instance, Daniel Drake (1816), Reuben D. Mussey (1809), George Hayward (1812), and Thomas Sewall (1811).
25. Thomas M. Doerflinger, *A Vigorous Spirit of Enterprise: Merchants and Economic Development in Revolutionary Philadelphia* (Chapel Hill: University of North Carolina Press, 1986), 5; Bruce Laurie, *Working People of Philadelphia, 1800–1850* (Philadelphia: Temple University Press, 1980), 9–10.
26. Susan E. Klepp, "Demography in Early Philadelphia, 1690–1860," *Proceedings of the American Philosophical Society* 133, no. 2 (1989): 85–111.
27. Thomas Sutton, *Tracts on Delirium Tremens, on Peritonitis, and on Some Other Internal Inflammatory Affections, and on the Gout* (London: Thomas Underwood, 1813).
28. Ron Roizen, "How Does the Nation's 'Alcohol Problem' Change from Era to Era? Stalking the Social Logic of Problem-Definition Transformations since Repeal," in *Altering American Consciousness: The History of Alcohol and Drug Use in the United States, 1800–2000*, ed. Sarah Tracy and Caroline Jean Acker (Amherst: University of Massachusetts Press, 2004).

CHAPTER ONE

1. Charles Brockden Brown, *Edgar Huntly, or Memoirs of a Sleepwalker*, vol. 4 of *Charles Brockden Brown's Novels* (1799, Port Washington, NY: Kennikat Press, 1963), 217–19.
2. Michael Meranze, *Laboratories of Virtue: Punishment, Revolution, and Authority in Philadelphia, 1760–1835* (Chapel Hill: University of North Carolina Press, 1996), 99–104.
3. Drew R. McCoy, *The Elusive Republic: Political Economy in Jeffersonian America* (Chapel Hill: University of North Carolina Press, 1980), 136–84.
4. David W. Conroy, *In Public Houses: Drink and the Revolution of Authority in Colonial Massachusetts* (Chapel Hill: University of North Carolina Press, 1995); Susanna Barrows and Robin Room, "Introduction," in *Drinking: Behavior and Belief in Modern History*, ed. Susanna Barrows and Robin Room (Berkeley: University of California Press, 1991); Peter Thompson, *Rum Punch and Revolution: Taverngoing and Public Life in Eighteenth-Century Philadelphia* (Philadelphia: University of Pennsylvania Press, 1999), 1–20.
5. Catherine O'Donnell Kaplan, *Men of Letters in the Early Republic* (Chapel Hill: University of North Carolina Press, 2008), 23–31; Sarah Knott, *Sensibility and the American Revolution* (Chapel Hill: University of North Carolina Press, 2009), chap. 5.
6. Conroy, *In Public Houses*, 310–22.
7. Simon Finger, *The Contagious City: The Politics of Public Health in Early Philadelphia* (Ithaca, NY: Cornell University Press, 2012), 111–12; W. J. Rorabaugh, *The Alcoholic Republic: An American Tradition* (New York: Oxford University Press, 1979), 38; Thompson, *Rum Punch and Revolution*, 203.
8. Benjamin Rush, *Medical Inquiries and Observations*, 2nd ed., 4 vols. (Philadelphia: J. Conrad, 1805), 1:iii, 335–84.

9. Thomas R. Pegram, *Battling Demon Rum: The Struggle for a Dry America, 1800–1933* (Chicago: Ivan R. Dee, 1998), 13–15; Rorabaugh, *Alcoholic Republic*, 40–48; Carl A. L. Binger, *Revolutionary Doctor: Benjamin Rush, 1746–1813* (New York: W. W. Norton, 1966), 197–201; James R. McIntosh, "Alcoholism," in *Alcohol and Temperance in Modern History: A Global Encyclopedia*, ed. Jack S. Blocker, David M. Fahey, and Ian Tyrrell (Santa Barbara, CA: ABC-Clio, 2003).
10. John Bell, *An Address to the Medical Students' Temperance Society of the University of Pennsylvania* (Philadelphia, 1833), 9; *The Anniversary Report of the Managers of the Pennsylvania Society for Discouraging the Use of Ardent Spirits* (Philadelphia: Pennsylvania Society for Discouraging the Use of Ardent Spirits, 1831), 68.
11. Harry G. Levine, "The Discovery of Addiction: Changing Conceptions of Habitual Drunkenness in America," *Journal of Studies on Alcohol* 39, no. 1 (1978): 143–74; Sarah Tracy and Caroline Jean Acker, eds., *Altering American Consciousness: The History of Alcohol and Drug Use in the United States, 1800–2000* (Boston: University of Massachusetts Press, 2004), 12; Sarah Tracy, *Alcoholism in America: From Reconstruction to Prohibition* (Baltimore: Johns Hopkins University Press, 2005), 2.
12. Peter Ferentzy, "From Sin to Disease: Differences and Similarities between Past and Current Conceptions of Chronic Drunkenness," *Contemporary Drug Problems* 28, no. 3 (2001): 363–275; Levine, "Discovery of Addiction," 143–71; Mariana Valverde, *Diseases of the Will: Alcohol and the Dilemmas of Freedom* (New York: Cambridge University Press, 1998); Jessica Warner, "Resolvd to Drink No More," *Journal of Studies on Alcohol* 55, no. 6 (1994): 685–91; William L. White, *Slaying the Dragon: The History of Addiction Treatment and Recovery in America* (Bloomington, IL: Chestnut Health Systems, 1998).
13. Robin Room, Thomas Babor, and Jürgen Rehm, "Alcohol and Public Health," *Lancet* 365 (2005): 519–30; Ron Roizen, "How Does the Nation's 'Alcohol Problem' Change from Era to Era? Stalking the Social Logic of Problem-Definition Transformations since Repeal," in *Altering American Consciousness: The History of Alcohol and Drug Use in the United States, 1800–2000*, ed. Sarah Tracy and Caroline Jean Acker (Amherst: University of Massachusetts Press, 2004).
14. Tracy, *Alcoholism in America*; Howard I. Kushner, "Taking Biology Seriously: The Next Task for Historians of Addiction?" *Bulletin of the History of Medicine* 80 (2006): 115–43.
15. Binger, *Revolutionary Doctor*, 281; George Hayward, "Some Observations on Dr. Rush's Work on the Diseases of the Mind," *New England Journal of Medicine and Surgery* 7 (1818): 18–34.
16. Gordon Wood, *The Radicalism of the American Revolution* (New York: Random House, 1991), 189–225.
17. Benjamin Rush, "Of the Mode of Education Proper in a Republic," quoted in Donald J. D'Elia, "Benjamin Rush: Philosopher of the American Revolution," *Transactions of the American Philosophical Society* 64, no. 5 (1974): 70.
18. Jessica Warner, *Craze: Gin and Debauchery in an Age of Reason* (New York: Four Walls Eight Windows, 2002), 3.
19. Roy Porter, "Introduction," in *An Essay Medical, Philosophical, and Chemical on Drunkenness, and Its Effects on the Human Body* by Thomas Trotter, ed. Roy Porter, Tavistock Classics in the History of Psychiatry (New York: Routledge, 1988), ix–xi.
20. Selwyn H. H. Carrington, "Sugar, Molasses, and Rum," in *History of World Trade since 1450*, ed. John J. McCusker (Detroit: Macmillan Reference USA, 2006); Mark Ed-

ward Lender and James Kirby Martin, *Drinking in America: A History* (New York: Free Press, 1982), 4–33.

21. Sarah Hand Meacham, *Every Home a Distillery: Alcohol, Gender, and Technology in the Colonial Chesapeake* (Baltimore: Johns Hopkins University Press, 2009), 6–23.
22. Alan Taylor, *American Colonies* (New York: Penguin, 2001), 208–11, 310–14.
23. John J. McCusker, "The Rum Trade and the Balance of Payments of the Thirteen Continental Colonies, 1650–1775," *Journal of Economic History* 30, no. 1 (1970): 24–27.
24. Cotton Mather, *Sober Considerations on a Growing Flood of Iniquity* (Boston: John Allen for Nicholas Boone, 1708), 2–3; Conroy, *In Public Houses*, 61–75.
25. Rorabaugh, *Alcoholic Republic*, 29.
26. Conroy, *In Public Houses*, 85–98, 154–56; Rorabaugh, *Alcoholic Republic*, 8–9; Lender and Martin, *Drinking in America*, 32–33; Daniel Walker Howe, *What Hath God Wrought: The Transformation of America, 1815–1848* (New York: Oxford University Press, 2007), 31.
27. Warner, "Resolvd to Drink No More," 688.
28. Conroy, *In Public Houses*, 49–56.
29. Increase Mather, *Wo to Drunkards: Two Sermons Testifying against the Sin of Drunkenness* (1673; Boston: Timothy Green, 1712), 7.
30. Ibid., i.
31. Conroy, *In Public Houses*, 96–98; Thompson, *Rum Punch and Revolution*, 33–51.
32. Peter Shaw, *The Juice of the Grape, or Wine Preferable to Water* (London: W. Lewis, 1724); David Hancock, *Oceans of Wine: Madeira and the Emergence of American Trade and Taste* (New Haven, CT: Yale University Press, 2009), 317–55.
33. Robert Burton, *The Anatomy of Melancholy: What It Is, with All the Kinds, Causes, Symptoms, Prognostics, and Several Cures of It* (London: Chatto and Windus, 1883), 146–47.
34. Michael MacDonald, *Mystical Bedlam: Madness, Anxiety, and Healing in Seventeenth-Century England* (New York: Cambridge University Press, 1973).
35. Warner, "Resolvd to Drink No More," 686; Ernest L. Abel, "Gin Lane: Did Hogarth Know about Fetal Alcohol Syndrome?" *Alcohol and Alcoholism* 36, no. 2 (2001): 131–34; Thomas Wilson, *Distilled Liquors the Bane of the Nation* (London: J. Roberts, 1736); Josiah Tucker, *An Impartial Enquiry into the Benefits and Damages Arising to the Nation from the Present Very Great Use of Low-Priced Spirituous Liquors* (London: T. Trye, 1751).
36. Quoted in Fiona Haslam, *From Hogarth to Rowlandson: Medicine in Art in Eighteenth Century Britain* (Liverpool: Liverpool University Press, 1997), 119.
37. George Cheyne, *An Essay of Health and Long Life*, 2nd ed. (London: G. Strahan, 1725), 52. Also see Thomas Short, *Vinum Britannicum, or An Essay on the Properties and Effects of Malt Liquors; Wherein Is Considered, in What Cases, and to What Constitutions, They Are Either Beneficial or Injurious* (London: D. Midwinter, 1727), 6; George Cheyne, *Discourses on Tea, Sugar, Milk, Made-Wines, Spirits, Punch, Tobacco, &c.* (London: T. Longman, 1750); William Cadogan, *A Dissertation on the Gout*, 10th ed. (Boston: Henry Knox, 1772), 16; William F. Bynum, "Chronic Alcoholism in the First Half of the Nineteenth Century," *Bulletin of the History of Medicine* 42, no. 2 (1968): 166–67; Warner, "Resolvd to Drink No More," 688.
38. George Cheyne, *The English Malady, or A Treatise of Nervous Diseases of All Kinds, as Spleen, Vapours, Lowness of Spirits, Hypochondriacal and Hysterical Distempers, Etc.* (1733; Delmar, NY: Scholars Facsimiles and Reprints, 1976), 35, quoted in

G. J. Barker-Benfield, *The Culture of Sensibility: Sex and Society in Eighteenth-Century Britain* (Chicago: University of Chicago Press, 1992), 12.

39. Ibid., 1–36.
40. [S. A. D.] Tissot, *Advice to the People in General, with regard to Their Health* (Edinburgh: A. Donaldson, 1766); William Buchan, *Domestic Medicine, or The Family Physician: Being an Attempt to Render the Medical Art More Generally Useful, by Shewing People What Is Their Own Power both with Respect to the Prevention and Cure of Diseases*, 2nd American ed. (Philadelphia: Joseph Crukshank, 1774); Abel, *"Gin Lane,"* 132.
41. Charles E. Rosenberg, "Medical Text and Social Context: Explaining William Buchan's *Domestic Medicine*," in *Explaining Epidemics and Other Studies in the History of Medicine* (New York: Cambridge University Press, 1992), 32–56; Knott, *Sensibility and the American Revolution*, 88–90.
42. Buchan, *Domestic Medicine*, 69–72.
43. Anthony Benezet, *The Mighty Destroyer Displayed: In Some Account of the Dreadful Havock Made by the Mistaken Use as Well as Abuse of Distilled Spirituous Liquors* (Philadelphia: Joseph Crukshank, 1774), 14–15.
44. For instance, ibid., 34–36.
45. Cheyne, *Essay of Health*, 52.
46. Buchan, *Domestic Medicine*, 72.
47. Benezet, *Mighty Destroyer Displayed*, 8.
48. Benjamin Rush, "An Inquiry into the Natural History of Medicine among the Indians of North-America and a Comparative View of Their Diseases and Remedies with Those of Civilized Nations," in *Medical Inquiries and Observations* (Philadelphia: Prichard and Hall, 1789), 73.
49. Peter Kafer, *Charles Brockden Brown's Revolution and the Birth of American Gothic* (Philadelphia: University of Pennsylvania Press, 2004), 181; Charles Brockden Brown, "Charles B. Brown to Bringhurst, July 29, August 1, 1793 (Bennett Census no. 48)," Charles Brockden Brown Collection, Bowdoin College Library, New Brunswick, ME.
50. Historians and literary scholars recognize Brown as a "touchstone for understanding the cultural transformations and conflicts of the early republic." See Philip Barnard, Mark L. Kamrath, and Stephen Shapiro, "Introduction," in *Revising Charles Brockden Brown: Culture, Politics, and Sexuality in the Early Republic*, ed. Philip Barnard, Mark L. Kamrath, and Stephen Shapiro (Knoxville: University of Tennessee Press, 2004), iii–xxi.
51. Karen Halttunen, "Humanitarianism and the Pornography of Pain in Anglo-American Culture," *American Historical Review* 100, no. 2 (1995): 305–11.
52. Kenneth Silverman, *A Cultural History of the American Revolution: Painting, Music, Literature and the Theatre in the Colonies and the United States from the Treaty of Paris to the Inauguration of George Washington, 1763–1789* (New York: Thomas Y. Crowell, 1976), 84–85, 141, 252–53; Knott, *Sensibility and the American Revolution*, 174.
53. Linda K. Kerber, *Women of the Republic: Intellect and Ideology in Revolutionary America* (Chapel Hill: University of North Carolina Press, 1980), 199–200.
54. Benjamin Rush, *An Inquiry into the Effects of Spirituous Liquors on the Human Body: To Which Is Added, a Moral and Physical Thermometer* (Boston: Thomas and Andrews, 1790), 12.
55. Benjamin Rush, *An Inquiry into the Effects of Ardent Spirits upon the Human Body and Mind, with an Account of the Means of Preventing, and of the Remedies for Curing Them*, 8th ed., ed. Gerald Grob, Addiction in America: Drug Abuse and Alcoholism (New York: Arno Press, 1981), 27.

56. For an overview of American drug panics, see Harry G. Levine and Craig Reinarman, "Crack in Context: America's Latest Demon Drug," in *Crack in America: Demon Drugs and Social Justice* (Berkeley: University of California Press, 1997).
57. Rorabaugh, *Alcoholic Republic*, 35; Conroy, *In Public Houses*, 265–68; Benjamin L. Carp, *Rebels Rising: Cities and the American Revolution* (New York: Oxford University Press, 2007).
58. Richard Stott, *Jolly Fellows: Male Milieus in Nineteenth-Century America* (Baltimore: Johns Hopkins University Press, 2009), 9.
59. Woody Holton, *Unruly Americans and the Origins of the Constitution* (New York: Hill and Wang, 2007), 145–61.
60. John K. Alexander, *Render Them Submissive: Responses to Poverty in Philadelphia, 1760–1800* (Amherst: University of Massachusetts Press, 1980), 61–85; Meranze, *Laboratories of Virtue*, 87–96.
61. David Freeman Hawke, *Benjamin Rush: Revolutionary Gadfly* (Indianapolis: Bobbs-Merrill, 1971), 300.
62. Quoted in ibid., 303.
63. Nina Reid-Maroney, *Philadelphia's Enlightenment, 1740–1800: Kingdom of Christ, Empire of Reason* (Westport, CT: Greenwood Press, 2001), 78; Knott, *Sensibility and the American Revolution*, 72–73; Henry May, *The Enlightenment in America* (New York: Oxford University Press, 1976), 197.
64. Hawke, *Benjamin Rush*, 180–279; Meranze, *Laboratories of Virtue*, 67; Alexander, *Render Them Submissive*, 6.
65. Hawke, *Benjamin Rush*, 49–50.
66. Rush, "An Address to the People of the United States," *American Museum* 1 (January 1787): 8–11. Quoted in D'Elia, "Benjamin Rush," 5.
67. Finger, *Contagious City*, 103–6.
68. Quoted in Sarah Knott, "Benjamin Rush's Ferment: Enlightenment Medicine and Republican Citizenship in Revolutionary America," presented at the McNeil Center for Early American Studies Seminar Series, November 5, 2004, 6.
69. Richard Harrison Shryock, "The Medical Reputation of Benjamin Rush: Contrasts over Two Centuries," *Bulletin of the History of Medicine* 45, no. 6 (1971): 507–52.
70. Christopher Lawrence, "The Nervous System and Society in the Scottish Enlightenment," in *Natural Order: Historical Studies of Scientific Culture*, ed. Barry Barnes and Steven Shapin (Beverly Hills, CA: Sage Publications, 1979), 19–40; William F. Bynum, *Science and the Practice of Medicine in the Nineteenth Century* (New York: Cambridge University Press, 1994), 11–18; Knott, *Sensibility and the American Revolution*, 74–82; Barker-Benfield, *Culture of Sensibility*, 1–36.
71. David Hume, *A Treatise of Human Nature*, Penguin Classics (New York: Penguin, 1969), 299–300; Anya Taylor, *Bacchus in Romantic England: Writers and Drink, 1780–1830* (New York: St. Martin's Press, 1999), 65.
72. Benjamin Rush, *Benjamin Rush's Lectures on the Mind*, ed. Eric T. Carlson, Jeffrey L. Wollock, and Patricia S. Noel (Philadelphia: American Philosophical Society, 1981), 83.
73. Ibid., 87.
74. Ibid., 88.
75. Ibid., 84.
76. Ibid., 238.
77. Lawrence, "Nervous System and Society in the Scottish Enlightenment," 32–33; Jessica Risken, *Science in the Age of Sensibility: The Sentimental Empiricists of the French Enlightenment* (Chicago: University of Chicago Press, 2002), 1–10.

78. Knott, *Sensibility and the American Revolution*, 201–17.
79. David Hartley, *Observations on Man, His Frame, His Duty, and His Expectations* (London: S. Richardson for James Leake and Wm. Frederick, 1749); Alan Richardson, *British Romanticism and the Science of the Mind* (New York: Cambridge University Press, 2001), 9; D'Elia, "Benjamin Rush," 24, 27, 68.
80. Rush's view of the mind resembled the new theories of phrenology and craniology circulating in Europe, but Rush said of Franz Joseph Gall's system, "It may be real, but it is by no means certain." Nevertheless, Rush's psychology posited a material basis for mental functions, just as Gall proposed to read human capacities in the structure of the skull, the crucial difference being that Rush could not demonstrate physical structures whereas Gall did. Rush, *Benjamin Rush's Lectures on the Mind*, 425–29.
81. Benjamin Rush, *Six Introductory Lectures, to Courses of Lectures, upon the Institutes and Practice of Medicine, Delivered in the University of Pennsylvania* (Philadelphia: John Conrad, 1801), 110–11.
82. Benjamin Rush, *An Oration Delivered before the American Philosophical Society, Containing an Enquiry into the Influence of Physical Causes on the Moral Faculty* (Philadelphia: Charles Cist, 1786), 40.
83. Ibid., 32; Kaplan, *Men of Letters*.
84. Rush, *Benjamin Rush's Lectures on the Mind*, 407.
85. D'Elia, "Benjamin Rush," 73; Knott, *Sensibility and the American Revolution*, 211–17.
86. Benjamin Rush, "Introduction to Lectures upon Apoplexy, Palsy, Coma and Epilepsy," Benjamin Rush Papers, Library Company of Philadelphia, 1811, 1–2.
87. Ibid., 2.
88. Ibid., 3.
89. Paul E. Kopperman, "Venerate the Lancet: Benjamin Rush's Yellow Fever Therapy in Context," *Bulletin of the History of Medicine* 78 (Fall 2004): 539–74.
90. Rush, "Introduction to Lectures upon Apoplexy, Palsy, Coma and Epilepsy," 17–18.
91. Rush, *Inquiry into the Effects of Spirituous Liquors*, 4–5.
92. Benjamin Rush, "Diseases and Disorders of the Liver," Benjamin Rush Papers, Library Company of Philadelphia, 2.
93. Terry Castle, *The Female Thermometer: Eighteenth-Century Culture and the Invention of the Uncanny* (New York: Oxford University Press, 1995), 21–43; Lisa Forman Cody, *Birthing the Nation: Sex, Science, and the Conception of Eighteenth-Century Britons* (New York: Oxford University Press, 2005), 143–44.
94. Clare A. Lyons, *Sex among the Rabble: An Intimate History of Gender and Power in the Age of Revolution; Philadelphia, 1730–1830* (Chapel Hill: University of North Carolina Press, 2006), 352–53; Knott, *Sensibility and the American Revolution*, 311–22; D'Elia, "Benjamin Rush," 102–5.
95. Benjamin Rush, *Letters of Benjamin Rush*, vol. 2, *1793–1813*, ed. L. H. Butterfield (Princeton, NJ: Princeton University Press, 1951), 976–79.
96. A biographer of Rush has interpreted this dream as evidence that Rush harbored frustrated political ambitions. Carl A. L. Binger, "The Dreams of Benjamin Rush," *American Journal of Psychiatry* 125, no. 12 (1969): 1656–58.
97. Benjamin Rush, "Of Drunkenness and Its Cures," 3, Benjamin Rush Papers, Library Company of Philadelphia. This lecture closely resembles the fourth edition of the *Inquiry*, published sometime between 1800 and 1805.
98. Ibid., 11.
99. Ibid., 20.

100. Benjamin Rush, "Facts and Documents on Moral Derangement as Exemplified Chiefly in Murder," Benjamin Rush Papers, Library Company of Philadelphia.
101. Benjamin Rush, *Observations on Diseases of the Mind* (Philadelphia: Kimber and Richardson, 1812), 267–68.
102. Roy Porter, *Mind-Forg'd Manacles: A History of Madness in England from the Restoration to the Regency* (London: Athlone Press, 1987), 174–97.
103. The earliest set of student notes on the diseases of the mind that I have been able to locate is dated February 1809. See Daniel J. Swiney Papers, 1808–1827, College of Physicians of Philadelphia.
104. Advertisements for medical books were common in Philadelphia newspapers, and these books were often advertised along with literary works, clearly meant to appeal to a literate, book-buying audience—not simply elite medical men. The *Zoonomia*, for instance, was regularly advertised this way. See *Poulson's American Daily Advertiser*, November 18, 1802; December 9, 1802; January 14, 1808.
105. Dora B. Weiner, "Mind and Body in the Clinic: Philippe Pinel, Alexander Crichton, Dominique Esquirol, and the Birth of Psychiatry," in *The Languages of Psyche: Mind and Body in Enlightenment Thought: Clark Library Lectures, 1985–1986*, ed. G. S. Rousseau (Berkeley: University of California Press, 1991).
106. Porter, *Mind-Forg'd Manacles*, 174–97.
107. Richardson, *British Romanticism and the Science of the Mind*, 1–38.
108. John Locke, quoted in Roy Porter, *Flesh in the Age of Reason* (New York: W. W. Norton, 2003), 311.
109. Philippe Pinel, *A Treatise on Insanity: Translated from the French by D. D. Davis* (New York: Hafner, 1962), 118; Jan Goldstein, *Console and Classify: The French Psychiatric Profession in the Nineteenth Century*, 2nd ed. (Chicago: University of Chicago Press, 2002), 65–119.
110. Alexander Crichton, *Inquiry into the Nature and Origin of Mental Derangement*, vol. 2 (Philadelphia: T. Cadell and W. Davies, 1798), 96. On Crichton's influence on Pinel, see Weiner, "Mind and Body in the Clinic."
111. Dror Wahrman, *The Making of the Modern Self: Identity and Culture in Eighteenth-Century England* (New Haven, CT: Yale University Press, 2004), 265–311.
112. The notes survive, but they are undated and contain many revisions. References in the notes demonstrate that Rush delivered this lecture several times in 1809 to 1812. Page numbers appear irregularly. Many notations and revisions are scribbled on the back or in the margins of previous drafts. In the following citations, page numbers are given when possible. Benjamin Rush, "Facts and Documents on Moral Derangement as Exemplified Chiefly in Murder," Benjamin Rush Papers, Library Company of Philadelphia.
113. Rush, "On Moral Derangement," 1.
114. Rush, "On Moral Derangement," 27.
115. Knott, *Sensibility and the American Revolution*, 311–12; Karen Halttunen, *Murder Most Foul: The Killer and the American Gothic Imagination* (Cambridge, MA: Harvard University Press, 1998), 1–6.
116. Halttunen, *Murder Most Foul*, 1–6, 33–59, 66–67.
117. Rush, *Letters of Benjamin Rush, 1793–1813*, 2:874–75.
118. Charles Brockden Brown, *Wieland, or The Transformation, an American Tale*, ed. Sydney J. Krause and S. W. Reid, Novels and Related Works of Charles Brockden Brown (Kent, OH: Kent State University Press, 1977), 3. Crichton's *Inquiry* was published

in Philadelphia the same year as *Wieland*. Crichton, *Inquiry into the Nature and Origin of Mental Derangement*, 2.

119. Larzer Ziff, "A Reading of *Wieland*," *Proceedings of the Modern Language Association* 77 (1962): 51–57; Peter Kafer, "Charles Brockden Brown and Revolutionary Philadelphia: An Imagination in Context," *Pennsylvania Magazine of History and Biography* 116, no. 4 (1992): 488.
120. Kafer, *Charles Brockden Brown's Revolution and the Birth of American Gothic*, 125; Brown, *Wieland*, 87.
121. Brown, *Wieland*, 87.
122. Rush, "On Moral Derangement," 22.
123. Ibid., 21.
124. Ibid., 25–26.
125. Ibid., 21.
126. Tracy, *Alcoholism in America*, 2.
127. Rush, *Observations on Diseases of the Mind*, 265–66.
128. *Aurora General Advertiser*, November 3, 1812, 1; Hayward, "Observations on Dr. Rush's Work," 33–34.
129. Charles Caldwell, *Autobiography of Charles Caldwell, M.D.* (Philadelphia: Lippincott, Grambo, 1855), 312–16; Weiner, "Mind and Body in the Clinic," 289–90.
130. Kopperman, "Venerate the Lancet," 569–70.
131. Knott, *Sensibility and the American Revolution*, 326–27.
132. Jennifer Ford, "Samuel Taylor Coleridge and the Pains of Sleep," *History Workshop Journal* 48 (1999): 169–86.
133. Rush, *Benjamin Rush's Lectures on the Mind*, 669.
134. Alethea Hayter, *Opium and the Romantic Imagination* (Berkeley: University of California Press, 1968), 67–83.
135. Richardson, *British Romanticism and the Science of the Mind*, 45–65.
136. Rush, *Benjamin Rush's Lectures on the Mind*, 652–64.
137. Thomas D. Mitchell, "Notes on the Lectures of Benj. Rush, Professor of the Institutes and Practice of Medicine in the University of Pennsylvania, 1809–1811," College of Physicians of Philadelphia, 1809–11, 309.
138. Ibid.; "Notes on the Lectures of Benjamin Rush," College of Physicians of Philadelphia, 1810?, 403–4.
139. Rush, *Benjamin Rush's Lectures on the Mind*, 450–52.
140. Ibid., 452.
141. *Aurora General Advertiser*, October 3, 1813, 3.
142. Richardson, *British Romanticism and the Science of the Mind*, 51.
143. David Brewster, *Letters on Natural Magic Addressed to Sir Walter Scott, Bart.* (New York: J. and J. Harper, 1832), 307–12.
144. *Aurora General Advertiser*, July 24, 1820, 3.
145. Rush, "On Moral Derangement."

CHAPTER TWO

1. Pliny Earle, "An Analysis of the Cases of Delirium Tremens Admitted into the Bloomingdale Asylum for the Insane from June 16th, 1821 to December 31st, 1844," *American Journal of the Medical Sciences* 15 (1848): 82–83.
2. Ibid., 83.
3. In Philadelphia, "mania a potu" and "delirium tremens" were the most common

terms. Doctors used them interchangeably throughout the antebellum period, sometimes for the same patient.

4. W. J. Rorabaugh, *The Alcoholic Republic: An American Tradition* (New York: Oxford University Press, 1979), 169–73. Similarly, Ric N. Caric has argued that the symptoms of delirium tremens were new, deriving from changing drinking patterns and cultural transformations caused by early industrialization. See Ric N. Caric, "The Man with the Poker Enters the Room: Delirium Tremens and Popular Culture in Philadelphia, 1828–1850," *Pennsylvania Magazine* 74, no. 4 (2007): 452–91.
5. Isaac Clarkson Snowden, "On Mania á Potu," *Eclectic Repertory and Analytical Review* 5 (1815): 372.
6. Ibid., 372–79.
7. Joseph Klapp, "Note to James Rush," Historical Society of Pennsylvania, Philadelphia; Joseph Klapp, "An Attempt to Point Out a New, and Successful Method of Treating Mania á Temulentia," *Eclectic Repertory and Analytical Review* 7 (1817): 251–64; Joseph Klapp, "On Temulent Diseases," *American Medical Recorder* 1 (1818): 462–78.
8. James Martin Staughton, "Observations on Mania a Potu" (MD diss., University of Pennsylvania, 1821), 8.
9. For instance, John Hook Griffin, "An Inaugural Dissertation on Mania a Potu" (MD diss., University of Pennsylvania, 1826), 12–13; Daniel Drake, "Observations on Temulent Diseases," *American Medical Recorder* 2 (1819): 60–65.
10. Simon Baatz, "'A Very Diffused Disposition': Dissecting Schools in Philadelphia, 1823–1825," *Pennsylvania Magazine of History and Biography* 108, no. 2 (1984): 206; Leonard K. Eaton, "Medicine in Philadelphia and Boston, 1805–1830," *Pennsylvania Magazine of History and Biography* 75, no. 1 (1951): 73; Daniel Kilbride, "Southern Medical Students in Philadelphia, 1800–1861: Science and Sociability in the 'Republic of Medicine,'" *Journal of Southern History* 65, no. 4 (1999): 697.
11. US Army, *Index-Catalogue of the Library of the Surgeon-General's Office*, 1st ser. (Washington, DC: Government Printing Office, 1880–95).
12. William F. Bynum, *Science and the Practice of Medicine in the Nineteenth Century* (New York: Cambridge University Press, 1994), 31–33; Lester S. King, *Transformations in American Medicine: From Benjamin Rush to William Ostler* (Baltimore: Johns Hopkins University Press, 1991), 72–78.
13. Erasmus Darwin, *Zoonomia, or The Laws of Organic Life in Three Parts*, 2nd American ed., vol. 2 (Boston: Thomas and Andrews, 1804), 262.
14. John Haslam, *Observations on Madness and Melancholy: Including Practical Remarks on Those Diseases; Together with Cases; and an Account of the Morbid Appearances on Dissection*, 2nd ed. (London: J. Callow, 1809), 144–47.
15. Benjamin Rush, *An Inquiry into the Effects of Ardent Spirits upon the Human Body and Mind, with an Account of the Means of Preventing, and of the Remedies for Curing Them*, 8th ed., ed. Gerald Grob, Addiction in America: Drug Abuse and Alcoholism (New York: Arno Press, 1981), 10.
16. Benjamin Rush, *Observations on Diseases of the Mind* (Philadelphia: Kimber and Richardson, 1812), 164.
17. Joseph Lee, "A Dissertation on Mania: Submitted to the Examination of the Trustees and Medical Faculty of the University of Pennsylvania, 1811" (MD diss., University of Pennsylvania, 1811).
18. Daniel Drake, *A Discourse on Intemperance: Delivered at Cincinnati, March 1, 1828* (Cincinnati: Looker and Reynolds, 1828), 85.

19. Robert Macnish, *The Anatomy of Drunkenness*, 4th ed. (Glasgow: W. R. McPhun, 1832), 157.
20. Benjamin H. Coates, "Observations on Delirium Tremens, or the Disease Improperly Called Mania a Potu," *North American Medical and Surgical Journal* 4, no. 7 (1827): 34. See also George Gibson, "An Essay on Delirium Tremens" (MD diss., University of Pennsylvania, 1837), 2.
21. Coates, "Observations on Delirium Tremens," 224–25.
22. Benjamin H. Coates, "Observations on Delirium Tremens" (1827?), Coates Reynell Collection, Historical Society of Pennsylvania, Philadelphia.
23. Charles Randolph, "An Essay upon Mania a Potu" (MD diss., University of Pennsylvania, 1824), 4.
24. Snowden, "On Mania á Potu," 373.
25. Stephen Brown, "Observations on Delirium Tremens, or the Delirium of Drunkards, with Cases," *American Medical Recorder* 5 (1822): 194.
26. Ibid., 194–95.
27. Richard Sexton, "The Treatment of Mania a Potu by the Spider's Web" (MD diss., University of Pennsylvania, 1826), 3; Randolph, "Mania a Potu," 8.
28. "Notes on the Lectures of Benjamin Rush" [1810?], 401–4, College of Physicians of Philadelphia.
29. Drake, "Observations on Temulent Diseases," 60.
30. Walter Channing, "Cases of 'Delirium Tremens,' or of 'a Peculiar Disease of Drunkards,'" *New England Journal of Medicine* 8 (1819): 16.
31. A Copy of a Course of Medical Lectures Delivered by Nathaniel Chapman MD Professor of the Institutes and Practice of Medicine and Clinical Practice of the University of Pennsylvania, College of Physicians of Philadelphia.
32. Randolph, "Mania a Potu," 7–8.
33. Staughton, "Observations on Mania a Potu," 3.
34. Terry Castle, *The Female Thermometer: 18th Century Culture and the Invention of the Uncanny* (New York: Oxford University Press, 1995), 168–89.
35. Philippe Pinel, *A Treatise on Insanity: Translated from the French by D. D. Davis* (New York: Hafner, 1962), 73–74.
36. Alexander Crichton, *Inquiry into the Nature and Origin of Mental Derangement* (Philadelphia: T. Cadell and W. Davies, 1798), 2:9.
37. Joseph Nancrede, "On Mania a Potu," *American Medical Recorder* 1 (1818): 480.
38. Quoted in Castle, *Female Thermometer*, 183; Crichton, *Inquiry into the Nature and Origin of Mental Derangement*, 2:9.
39. Thomas D. Mitchell, "Notes on the Lectures of Benj. Rush, Professor of the Institutes and Practice of Medicine in the University of Pennsylvania, 1809–1811," College of Physicians of Philadelphia, Philadelphia, 1809–1811, 309.
40. Pennsylvania Hospital Archives, Collection of Cases, 1804–1828, American Philosophical Society, Philadelphia, 78.
41. Ibid., 61–65, 78, 80–83, 314, 332.
42. Samuel Burton Pearson, "Observations on Brain Fever," *Edinburgh Medical and Surgical Journal* 9 (1813): 330.
43. Snowden, "On Mania á Potu," 373.
44. James Carter, "Observations on Mania a Potu," *American Journal of the Medical Sciences* 6 (1830): 321.
45. Drake, "Observations on Temulent Diseases," 61.
46. Klapp, "Treating Mania á Temulentia," 256.

47. Ibid., 257–60.
48. Peter Kafer, *Charles Brockden Brown's Revolution and the Birth of American Gothic* (Philadelphia: University of Pennsylvania Press, 2004), xi–xxi.
49. Matthew Gregory Lewis, *"The Castle Spectre": A Drama in Five Acts* (Philadelphia: J. Bioren, 1801); David Grimsted, *Melodrama Unveiled: American Theater and Culture, 1800–1850*, new foreword by Lawrence W. Levine, Approaches to American Culture (Chicago: University of Chicago Press, 1968; repr. Berkeley: University of California Press, 1987). "The New Theatre," *Aurora General Advertiser* January 16, 1810, 3; September 21, 1811, 3; March 1, 1817, 3.
50. Richard D. Altick, *The Shows of London* (Cambridge, MA: Harvard University Press, 1978), 117–19.
51. Castle, *Female Thermometer*, 150–51; David Brewster, *Letters on Natural Magic Addressed to Sir Walter Scott, Bart.* (New York: J. and J. Harper, 1832), 77–87.
52. Altick, *Shows of London*, 217–20.
53. James W. Cook, *The Arts of Deception: Playing with Fraud in the Age of Barnum* (Cambridge, MA: Harvard University Press, 2001), 177.
54. Étienne-Gaspard Robertson, *Mémoires recreatifs, scientifiques et anecdotiques*, cited in Robert Eskind, "The Magic Lantern," in *The Magic Lantern of Dr. Thomas Kirkbride: A Catalog of the Lantern Slides, Stereo Views and Cartes-de-Visite at the Institute of Pennsylvania Hospital*, ed. Frances Gage and Carolyn Harper (Philadelphia: Atwater Kent Museum, 1992), 9.
55. Brewster, *Letters on Natural Magic*, 82.
56. *Aurora General Advertiser*, January 13, 1809, 3. Rubens and Raphaelle Peale likely first saw the phantasmagoria show in London in 1802. See editors' note, Charles Willson Peale, *The Belfield Farm Years*, ed. Sidney Hart and Toby A. Appel, vol. 3 of *The Selected Papers of Charles Willson Peale and His Family* (New Haven, CT: Yale University Press, 1988), 814.
57. Peale, *Belfield Farm Years*, 812–13; Rubens Peale, "Letterbooks," January 9, 1815, Peale Papers, American Philosophical Society, Philadelphia.
58. Earle, "Cases of Delirium Tremens"; Daniel Drake, *Analytical Report of a Series of Experiments in Mesmeric Somniloquism* (Louisville: F. W. Prescott, 1844), 49–50.
59. Coates, "Observations on Delirium Tremens," 38; John Alderson, "On Apparitions," *Edinburgh Medical and Surgical Journal* 6 (1810): 287–95; Pearson, "Observations on Brain Fever."
60. Archives of the Pennsylvania Hospital, Library Register, 1824—42, American Philosophical Society, Philadelphia.
61. Samuel Hibbert, *Sketches of the Philosophy of Apparitions, or An Attempt to Trace Such Illusions to Their Physical Causes*, ed. Robert L. Morris, Perspectives in Psychical Research (1824; New York: Arno Press, 1975), iii, 59–69, 111, 67.
62. Ibid., iii.
63. "Philadelphia Theatre," *Aurora General Advertiser*, March 1, 1817; "Walnut Street Theatre, January 13th, 1821," Playbills pre-1825, Historical Society of Pennsylvania, Philadelphia. This is one example of a much larger shift in how audiences viewed Hamlet. About 1800, for instance, new critics began to describe the character as being in possession of a complex psychological interiority. Margreta de Grazia, *Hamlet without Hamlet* (New York: Cambridge University Press, 2007).
64. Michael O'Connell, *Robert Burton* (Boston: Twayne, 1986), x; John Ferriar, *An Essay toward a Theory of Apparitions* (London: J. and J. Haddock, Warrington, 1812); "Comments on Sterne by John Ferriar, MD," *South Carolina State Gazette and Timo-*

thy's Daily Adviser, November 14, 1794; Robert Gordon Hallwachs, *The Vogue of Robert Burton, 1798–1832* (Urbana: University of Illinois Press, 1937).

65. Rush, *Observations on Diseases of the Mind*, 40, 119; *Catalogue of the Late Dr. Harvey Klapp, Dec'd: to Be Sold without Reserve* (Philadelphia: Carey, 1832); Archives of the Pennsylvania Hospital, Library Register, 1824—12, American Philosophical Society, Philadelphia.
66. Hibbert, *Philosophy of Apparitions*, 32, 40, 80, 235–42.
67. Ibid., 33.
68. Castle, *Female Thermometer*, 174.
69. John H. Gibbon, *Desultory Notes on the Origin, Uses, and Effects of Ardent Spirit* (Philadelphia: Adam Waldie, 1834), 76.
70. "Opiologia, or *Confessions of an English Opium-Eater*," *American Medical Recorder* 5, no. 3 (1822): 553.
71. I discuss this persistence in the epilogue.
72. Michael Sappol, *A Traffic of Dead Bodies: Anatomy and Embodied Social Identity in Nineteenth-Century America* (Princeton, NJ: Princeton University Press, 2002), 59.
73. Pearson, "Observations on Brain Fever," 326.
74. Snowden, "On Mania á Potu," 373.
75. Coates, "Observations on Delirium Tremens," 37.
76. John Godman, *Contributions to Physiological and Pathological Anatomy* (Philadelphia: H. C. Carey and I. Lea, 1825); James Webster, *Facts concerning Anatomical Instruction in Philadelphia* (Philadelphia: James Webster, 1832); Baatz, "'Very Diffused Disposition,'" 203–10; Benjamin H. Coates, "Lecture Notes for an Address Delivered at Opening the Winter Course of the Medical Lyceum of Philadelphia for 1825–1826" (1825), Coates Reynell Collection, Historical Society of Pennsylvania, Philadelphia, 1825.
77. John Godman, *Anatomical Investigations Comprising Descriptions of Various Fasciae of the Human Body, to Which Is Added an Account of Some Irregularities of Structure and Morbid Anatomy* (Philadelphia: H. C. Carey and I. Lea, 1824).
78. Matthew Baillie, *The Morbid Anatomy of the Most Important Parts of the Human Body* (Albany, NY: Thomas Spencer, 1795), i.
79. Anthony Benezet, *Remarks on the Nature and Bad Effects of Spirituous Liquors* (Philadelphia, 1775), 1.
80. James Staughton, "Observations on Mania a Potu," *Philadelphia Journal of the Medical and Physical Sciences* 3 (1821): 240.
81. For instance, Robert Ewing Kerr, "Mania a Potu" (MD diss., University of Pennsylvania, 1825), 12; William Collins, "Mania a Potu" (MD diss., University of Pennsylvania, 1828), 6–7.
82. Baatz, "'Very Diffused Disposition,'" 203–11.
83. Jacob Sharpless, "An Inaugural Dissertation on Mania a Temulenta, Submitted to the Examination of the Provost and Medical Professors in the University of Pennsylvania," in Two Essays 1816 and 1817, College of Physicians of Philadelphia, 2.
84. For instance, Charles Hufferable, "An Inaugural Dissertation on Dyspepsia" (MD diss., University of Pennsylvania, 1829); Benjamin W. Blackwood, "On the Derangement of the Digestive Organs" (MD diss., University of Pennsylvania, 1828); Ezra Stiles Meigs, "An Inaugural Dissertation on Hypochondriasis" (MD diss., University of Pennsylvania, 1822); Thomas J. Charlton, "An Inaugural Essay on Melancholia and Hypochondriasis" (MD diss., University of Pennsylvania, 1827).
85. Blackwood, "On the Derangement of the Digestive Organs," 13.

86. "An Essay on the Gastric Pathology of Insanity, and Certain Disorders of the Animal Functions," *American Medical Recorder* 1, no. 3 (1818): 3.
87. Gilbert Flagler, "A Case of Mania," *American Medical Recorder* 2 (1819): 185. Also see Sharpless, "Inaugural Dissertation."
88. John Harley Warner, *Against the Spirit of the System: The French Impulse in Nineteenth-Century American Medicine* (Princeton, NJ: Princeton University Press, 1998).
89. Samuel Jackson, "Clinical Reports of Cases Treated in the Infirmary of the Alms-House of the City and County of Philadelphia," *American Journal of the Medical Sciences* 1 (1827): 94.
90. Godman, *Contributions to Physiological and Pathological Anatomy*, 5–10.
91. John Harley Warner, "Remembering Paris: Memory and the American Disciples of French Medicine in the Nineteenth Century," *Bulletin of the History of Medicine* 65, no. 3 (1991): 301–25.
92. Thomas Sewall, *An Eulogy on Dr. Godman* (Washington, DC: Columbian College, 1830); Baatz, "'Very Diffused Disposition,'" 203–5; Samuel Jackson of Northumberland to Benjamin H. Coates, April 9, 1827, Coates Reynell Collection, Historical Society of Pennsylvania, Philadelphia; Jackson, "Clinical Reports," 93; William E. Horner, *A Treatise on Pathological Anatomy* (Philadelphia: Carey, Lea and Carey, 1829), xxiv–xxvi; Nancy Tomes, *A Generous Confidence: Thomas Story Kirkbride and the Art of Asylum-Keeping, 1840–1883* (New York: Cambridge University Press, 1984), 68. A clarification: Samuel Jackson and Samuel Jackson of Northumberland were different physicians. Both were members of various medical organizations in antebellum Philadelphia.
93. Webster, *Facts concerning Anatomical Instruction*.
94. William E. Horner, "Lectures on Physiology by Benjamin Rush," 1812, manuscript lecture notes, College of Physicians of Philadelphia.
95. Horner, *Treatise on Pathological Anatomy*, xxv–xvi.
96. F. J. V. Broussais, *History of Chronic Phlegmasiae, or Inflammations, Founded on Clinical Experience and Pathological Anatomy* (Philadelphia: Carey and Lea, 1831), 13.
97. Thomas Sewall, *The Pathology of Drunkenness, or The Physical Effects of Alcoholic Drinks, with Drawings of the Drunkard's Stomach* (Albany, NY: C. Van Benthuysen, 1841).
98. Benjamin Rush, *Medical Inquiries and Observations*, 2nd ed., 4 vols. (Philadelphia: J. Conrad, 1805), 1:401.
99. Horner, Lectures on Physiology by Benjamin Rush; Ruth Richardson, *Death, Dissection, and the Destitute*, 2nd ed. (Chicago: University of Chicago Press, 2000), 39.
100. Editors' introduction, in Benjamin Rush, *Benjamin Rush's Lectures on the Mind*, ed. Eric T. Carlson, Jeffrey L. Wollock, and Patricia S. Noel (Philadelphia: American Philosophical Society, 1981), 53.
101. Charles Caldwell, *Autobiography of Charles Caldwell, M.D.* (Philadelphia: Lippincott, Grambo, 1855), 313–16; George Hayward, "Some Observations on Dr. Rush's Work on the Diseases of the Mind," *New England Journal of Medicine and Surgery* 7 (1818): 18–34; Jackson, "Clinical Reports," 93.
102. Jackson, "Clinical Reports," 94.
103. Ibid., 94–95. In his 1805 MD dissertation, Joseph Klapp revealed himself as a critic of Rush and an early disciple of Bichat. Joseph Klapp, *A Chemico-physiological Essay, Disproving the Existence of an Aeriform Function in the Skin, and Pointing Out, by Experiment, the Impropriety of Ascribing Absorption to the External Surface of the Human Body* (Philadelphia: Joseph Klapp, 1805).

104. On American criticism of Broussais in the 1830s, see Warner, *Against the Spirit of the System*, 179–80.
105. Samuel Annan, "Report of Cases Treated in the Baltimore Almshouse Hospital," *American Journal of the Medical Sciences* 24 (1839–40): 325–26.
106. Ibid., 327–28.
107. Samuel Jackson of Northumberland to Benjamin H. Coates, May 3, 1827, Coates Reynell Collection, Historical Society of Pennsylvania, Philadelphia.
108. Coates, "Observations on Delirium Tremens," 48–49.
109. "Report of the Philadelphia Medical Society, on the Removal of the Alms-House Infirmary," *Register of Pennsylvania* 3, no. 16 (1829): 251–52; Lisa Rosner, "Thistle on the Delaware: Edinburgh Medical Education and Philadelphia Practice, 1800–1825," *Social History of Medicine* 5 (1992): 19–42.
110. William Barlow and David O. Powell, "A Dedicated Medical Student: Solomon Mordecai, 1819–1822," *Journal of the Early Republic* 7, no. 4 (1987): 382.
111. For a breakdown of states of origins of graduates of the University of Pennsylvania medical school, see Kilbride, "Southern Medical Students in Philadelphia," 703. Kilbride demonstrates that most graduates were from southern states. Note, however, that most medical students chose not to formally graduate from the university. It is not clear that most students in attendance were from the South.
112. Priscilla Ferguson Clement, *Welfare and the Poor in the Nineteenth-Century City: Philadelphia, 1800–1854* (Rutherford, NJ: Fairleigh Dickinson University Press, 1985), 83.
113. Guardians of the Poor, Overseers of the Poor, Minutes 1801–10, reel #947120, R.S. 35.6, Philadelphia City Archives, p. 261; "Books in the Possession of the Medical Department May 1st, 1810," Board of Physicians, Minutes, 1809–45, Guardians of the Poor, Record no. 35.39, Philadelphia City Archives.
114. *The Philanthropist, or Institutions of Benevolence, by a Pennsylvanian* (Philadelphia: Isaac Peirce, 1813), 90–94.
115. Charles E. Rosenberg, *The Care of Strangers: The Rise of America's Hospital System* (New York: Basic Books, 1987), 63; "Report of the Philadelphia Medical Society, on the Removal of the Alms-House Infirmary"; Charles Lawrence, *History of the Philadelphia Almshouses and Hospitals from the Beginning of the Eighteenth to the Ending of the Nineteenth Century, Covering a Period of Nearly Two Hundred Years . . .* (Philadelphia: Charles Lawrence, 1905), 160–62.
116. Warner, *Against the Spirit of the System*, 17–31.
117. William G. Rothstein, *American Medical Schools and the Practice of Medicine: A History* (New York: Oxford University Press, 1987), 33.
118. "Address Delivered at Opening the Winter Course of the Medical Lyceum of Philadelphia for 1825–1826," Benjamin Coates Medical Essays, Coates Reynell Collection, Historical Society of Pennsylvania, Philadelphia.
119. Joseph F. Kett, *The Formation of the American Medical Profession: The Role of Institutions, 1780–1860* (Westport, CT: Greenwood Press, 1980), 14–30.
120. Warner, *Against the Spirit of the System*, 17–18.
121. Baatz, "'Very Diffused Disposition,'" 206–11; Coates, "Lecture Notes for an Address Delivered at Opening the Winter Course of the Medical Lyceum."
122. Warner, *Against the Spirit of the System*, 125; Kilbride, "Southern Medical Students in Philadelphia," 709; Steven M. Stowe, *Doctoring the South: Southern Physicians and Everyday Medicine in the Mid-Nineteenth Century* (Chapel Hill: University of North Carolina Press, 2004), 59–68.

123. Godman, *Anatomical Investigations,* vii.
124. Ibid., vi.
125. Sappol, *Traffic of Dead Bodies,* 2.
126. Ibid., 96. On the practice of anatomy in Britain, see Richardson, *Death, Dissection, and the Destitute.*
127. Sappol, *Traffic of Dead Bodies,* 114–15.
128. Barlow and Powell, "Dedicated Medical Student," 387; Warner, *Against the Spirit of the System,* 25–28.
129. *Report of the Commissioners of the Penal Code, with the Accompanying Documents* (Harrisburg: SC: Stambaugh, 1828), 104.
130. "Notice Intended for the Report of the Commission on Penal Code, on Law against Dissection, 6/31/28," Benjamin Coates Papers—Medical Essays, etc., Coates Reynell Collection, Historical Society of Pennsylvania, Philadelphia.
131. Webster, *Facts concerning Anatomical Instruction;* Sappol, *Traffic of Dead Bodies,* 112–17. This problem was evident in European centers of medicine as well. See Richardson, *Death, Dissection, and the Destitute,* 52–72.
132. "A graveyard a short distance west . . . ," *Aurora General Advertiser,* December 28, 1826, 2.
133. "Burial Ground," *Saturday Bulletin,* October 1, 1831, 3.
134. Richardson, *Death, Dissection, and the Destitute,* 80, 272–75.
135. "The Night Hawk," *Mechanics' Free Press,* December 6, 1828, 1.
136. *"Disturbing the Dead!—Awful Disclosures!!" Public Ledger,* April 4, 1836, 2.
137. Richardson, *Death, Dissection, and the Destitute,* 30–31; Stowe, *Doctoring the South,* 62.
138. Barlow and Powell, "Dedicated Medical Student," 387.
139. "Medical Education," *New York Medical Inquirer* 1 (1830): 130, quoted in Sappol, *Traffic of Dead Bodies,* 80.
140. Ibid., 81.
141. Baatz, "'Very Diffused Disposition,'" 207.
142. Barlow and Powell, "Dedicated Medical Student," 3; Paul Starr, *The Social Transformation of American Medicine* (New York: Basic Books, 1982), 63.
143. A Copy of a Course of Medical Lectures Delivered by Nathaniel Chapman MD Professor of the Institutes and Practice of Medicine and Clinical Practice of the University of Pennsylvania, College of Physicians of Philadelphia, 148–49; Nathaniel Chapman, Dr. Chapman's Notes, College of Physicians of Philadelphia, 247–48; W. W. Gerhard, "Lectures on Clinical Medicine, Delivered at the Philadelphia Medical Institute: Delirium Tremens," *Medical Examiner* 1, no. 14 (1838): 223–24.
144. This remembrance of Joseph Klapp came in a memorial written for Joseph's son, William, who also worked as a physician in Philadelphia. David Francis Condie, *A Memoir of William H. Klapp, M.D., Read before the College of Physicians of Philadelphia, January 7, 1857* (Philadelphia: College of Physicians of Philadelphia, 1857), 5–6.
145. "Medical Societies," *Medical and Surgical Reporter* 1 (1859): 409.
146. Joseph Klapp, Collection of Documents concerning Property on West Second Street, Society Collection, Historical Society of Pennsylvania. A year later, in a further indication of his social aspirations, he purchased a $200 piano for his two young daughters. Joseph Klapp, Receipt Book, 1817–24, Historical Society of Pennsylvania.
147. Sappol, *Traffic of Dead Bodies,* 63–70; Kilbride, "Southern Medical Students in Philadelphia," 697–732.
148. Kilbride, "Southern Medical Students in Philadelphia," 699, 721–23. On the impor-

tance of "mental culture" to early nineteenth-century ideals of gentility, see Richard Bushman, *The Refinement of America: Persons, Houses, Cities* (New York: Vintage Books, 1993), 282–87.

149. Sharpless, "Inaugural Dissertation," 4.
150. Randolph, "Mania a Potu," 1.
151. Channing, "Cases of 'Delirium Tremens,'" 15.
152. Nancrede, "On Mania a Potu," 485.
153. In the dissertation, this passage is in quotation marks. I was unable to locate the original source. It may be that the student is quoting a lecture by a prominent teacher, perhaps Klapp. Staughton, "Observations on Mania a Potu," 246.
154. Brown, "Observations on Delirium Tremens," 195–96.
155. Bushman, *Refinement of America*, 282–87.

CHAPTER THREE

1. James Oldden Jr., Diary 1800–1801, 1806, 1824, Historical Society of Pennsylvania, 4–5, 11, 40–45.
2. Ibid., 271.
3. Ibid., 272.
4. Correspondence, "To James Oldden, Philadelphia, from John Quincy Adams," April 6, 1829, Historical Society of Pennsylvania.
5. Patient Admission Records, 1823–32, Pennsylvania Hospital Archives.
6. James Oldden, Register of Wills, administration file 126, book O, p. 19.
7. Stephen Simpson, *Biography of Stephen Girard, with His Will Affixed; Comprising an Account of His Private Life, Habits, Genius, and Manners* (Philadelphia: Thomas L. Bonsal, 1832); Bruce Dorsey, *Reforming Men and Women: Gender in the Antebellum City* (Ithaca, NY: Cornell University Press, 2002), 105.
8. Roger Davis, "An Inaugural Dissertation on Mania a Potu" (MD diss., University of Pennsylvania, 1827), 1; John Hook Griffin, "An Inaugural Dissertation on Mania a Potu" (MD diss., University of Pennsylvania, 1826), 1–2.
9. Thomas M. Doerflinger, *A Vigorous Spirit of Enterprise: Merchants and Economic Development in Revolutionary Philadelphia* (Chapel Hill: University of North Carolina Press, 1986), 342–44.
10. J. David Lehman, "Explaining Hard Times: Political Economy and the Panic of 1819 in Philadelphia" (PhD diss., University of California, Los Angeles, 1992), 1–5; Richard Stott, *Jolly Fellows: Male Milieus in Nineteenth-Century America* (Baltimore: Johns Hopkins University Press, 2009), 62–63.
11. Gary B. Nash, "Poverty and Politics in Early American History," in *Down and Out in Early America*, ed. Billy G. Smith (University Park: Pennsylvania State University Press, 2004), 7–9.
12. Lehman, "Explaining Hard Times," 305–7.
13. Nash, "Poverty and Politics," 17.
14. Charles Lawrence, *History of the Philadelphia Almshouses and Hospitals from the Beginning of the Eighteenth to the Ending of the Nineteenth Century Covering a Period of Nearly Two Hundred Years . . .* (Philadelphia: Charles Lawrence, 1905), 24.
15. Gary B. Nash, "Poverty and Poor Relief in Pre-Revolutionary Philadelphia," *William and Mary Quarterly* 33, no. 1 (1976): 9; Charles E. Rosenberg, *The Care of Strangers: The Rise of America's Hospital System* (New York: Basic Books, 1987), 22–25.
16. Simon Finger, *The Contagious City: The Politics of Public Health in Early Philadelphia* (Ithaca, NY: Cornell University Press, 2012), 115–16.

17. Nash, "Poverty and Poor Relief in Pre-Revolutionary Philadelphia," 28–30.
18. John K. Alexander, *Render Them Submissive: Responses to Poverty in Philadelphia, 1760–1800* (Amherst: University of Massachusetts Press, 1980), 52.
19. Pennsylvania Society for the Promotion of Public Economy, *Report of the Library Committee of the Pennsylvania Society for the Promotion of Public Economy, Containing a Summary of the Information Communicated by Sundry Citizens* (Philadelphia: Pennsylvania Society for the Promotion of Public Economy, 1817), 3.
20. Ibid., 19–20.
21. Ibid., 24.
22. Dorsey, *Reforming Men and Women*, 56–60; Michael Meranze, *Laboratories of Virtue: Punishment, Revolution, and Authority in Philadelphia, 1760–1835* (Chapel Hill: University of North Carolina Press, 1996), 229–30; Susan G. Davis, *Parades and Power: Street Theatre in Nineteenth-Century Philadelphia* (Philadelphia: Temple University Press, 1986), 24–25.
23. Lehman, "Explaining Hard Times," 260; Edward Balleisen, *Navigating Failure: Bankruptcy and Commercial Society in Antebellum America* (Chapel Hill: University of North Carolina Press, 2001), 25–48; Robert M. Blackson, "Pennsylvania Banks and the Panic of 1819: A Reinterpretation," *Journal of the Early Republic* 9, no. 3 (1989): 335–58.
24. *Aurora General Advertiser*, May 15, 1819, 2.
25. The report was printed in the *Aurora General Advertiser*, October 4, 1819, 2; Lehman, "Explaining Hard Times," 288.
26. Charles Sellers, *The Market Revolution: Jacksonian America, 1815–1846* (New York: Oxford University Press, 1991), 137, 48.
27. Lehman, "Explaining Hard Times," 307. For similar developments in Baltimore, see Seth Rockman, *Scraping By: Wage Labor, Slavery, and Survival in Early Baltimore*, ed. Cathy Matson, Studies in Early American Economy and Society from the Library Company of Philadelphia (Baltimore: Johns Hopkins University Press, 2009), 199.
28. Priscilla Ferguson Clement, "The Philadelphia Welfare Crisis of the 1820s," *Pennsylvania Magazine of History and Biography* 110 (1981): 150–65.
29. Lehman, "Explaining Hard Times," 1–32; Sellers, *Market Revolution*, 137, 48; Scott A. Sandage, *Born Losers: A History of Failure in America* (Cambridge, MA: Harvard University Press, 2005), 23.
30. Sellers, *Market Revolution*, 137, 48.
31. *Aurora General Advertiser*, July 27, 1819, 2.
32. *Aurora General Advertiser*, May 20, 1819, 2.
33. *Aurora General Advertiser*, May 19, 1819, 2.
34. William Milnor, "Mania a Temulentia" (MD diss., University of Pennsylvania, 1823), 1.
35. Griffin, "Inaugural Dissertation on Mania a Potu," 3.
36. Jacob Sharpless, "An Inaugural Dissertation on Mania a Temulenta, Submitted to the Examination of the Provost and Medical Professors in the University of Pennsylvania," in *Two Essays, 1816 and 1817* (Philadelphia: College of Physicians of Philadelphia, 1817), 3.
37. Charles Randolph, "An Essay upon Mania a Potu" (MD diss., University of Pennsylvania, 1824).
38. Milnor, "Mania a Temulentia," 19.
39. Balleisen, *Navigating Failure*, 6–21, 108–21.
40. Dorsey, *Reforming Men and Women*, 105; E. Anthony Rotundo, *American Manhood:*

Transformations in Masculinity from the Revolution to the Modern Era (New York: Basic Books, 1993), 1–30; Brian P. Luskey, *On the Make: Clerks and the Quest for Capital in Nineteenth-Century America* (New York: New York University Press, 2010), 21–53.

41. Karen Halttunen, *Confidence Men and Painted Women: A Study of Middle-Class Culture in America, 1830–1870* (New Haven, CT: Yale University Press, 1982), 1–55; Rodney Hessinger, *Seduced, Abandoned, and Reborn: Visions of Youth in Middle-Class America, 1780–1850* (Philadelphia: University of Pennsylvania Press, 2006), 148–76.
42. The following mortality statistics published by the Board of Health have been collected in Susan Klepp, *"The Swift Progress of Population": A Documentary and Bibliographic Study of Philadelphia's Growth, 1642–1859* (Philadelphia: American Philosophical Society, 1991).
43. The term mania a potu was first published in almshouse statistics; *Aurora General Advertiser*, October 31, 1814, 3.
44. As measured by the Board of Health, overall mortality in Philadelphia rose from 2,271 in 1808 to well over 5,000 annually throughout the 1830s; Susan E. Klepp, "Demography in Early Philadelphia, 1690–1860," *Proceedings of the American Philosophical Society* 133, no. 2 (1989): 85–111.
45. "Insanity" and "mania" appear inconsistently in the Board of Health's statements of deaths. Since both are diagnoses describing mental illness, figure 4 combines them.
46. "Mania-a-Potu," *Philadelphia Gazette*, reprinted in the *New-Hampshire Gazette*, September 7, 1824, 1.
47. This statistic is far higher than twenty-first century estimates of alcohol-related mortality. Robin Room, Thomas Babor, and Jürgen Rehm, "Alcohol and Public Health," *Lancet* 365 (2005): 519–30; *Report of the Committee, Appointed by the Philadelphia Medical Society, January 24, 1829, to Take into Consideration the Propriety of That Society Expressing Their Opinion with regard to the Use of Ardent Spirits* (Philadelphia: John Clarke, 1829), 5–6.
48. W. W. Gerhard, "Philadelphia Hospital," *Medical Examiner* 3, no. 3 (1840): 1850; John Prosser Tabb, "Statistics of the Causes of Death in the Philadelphia Hospital, Blockley, during a Period of Twelve Years," *American Journal of the Medical Sciences* 16 (1844): 365.
49. Patient records that might have substantiated Gerhard's claims have not survived. Records at the Pennsylvania Hospital conflict with both Gerhard's claims and the statistics in figure 4. "Medical Societies," *Medical and Surgical Reporter* 1 (1859): 409–11.
50. Patient Admission and Discharge, 1830–1850, Pennsylvania Hospital Historical Archives, Philadelphia.
51. Philadelphia Board of Health, "Registration of Deaths, Cemetery Returns, 1803–1860," on microfilm at the Historical Society of Pennsylvania, forms the basis of this study. Hereafter, Cemetery Returns HSP.
52. The cemetery returns sample comprises 57 percent of the deaths by delirium tremens and 70 percent of the deaths by intemperance reported by the board during the same years.
53. The survey found 398 women. Cemetery Returns HSP.
54. In the cemetery returns sample, 86 percent of delirium tremens victims were male and 45 percent of intemperance victims were women. Cemetery Returns HSP.
55. "Mania-a-Potu," *Philadelphia Gazette*, reprinted in the *New-Hampshire Gazette*, September 7, 1824, 1.
56. Northern Dispensary of Philadelphia for the Medical Relief of the Poor, Register of Patients, 1816–1862, Historical Society of Pennsylvania.

57. Egbert Guernsey, *Homoeopathic Domestic Practice: Containing Also Chapters on Anatomy, Physiology, Hygiene, and an Abridged Materia Medica* (New York: William Radde, 1853), 365.
58. "Various Statements and Estimates to Show the Annual Destruction of Health, Reason, and Life in the United States, Produced by the Use of Spirituous Liquors," in *Annual Report of the Executive Committee of the American Society for the Promotion of Temperance for the Year Ending November 1827* (Andover, MA, 1827), 61.
59. On anxieties related to female drunkenness, see Scott C. Martin, *Devil of the Domestic Sphere: Temperance, Gender, and Middle-Class Ideology, 1800–1860* (DeKalb: Northern Illinois University Press, 2008), 15–38.
60. Modern sociological studies have shown that consumption of alcohol and other drugs is often influenced by price and availability. Room, Babor, and Rehm, "Alcohol and Public Health"; W. J. Rorabaugh, "Alcohol in America," *OAH Magazine of History* 6, no. 2 (1991): 18.
61. Simon P. Newman, *Embodied History: The Lives of the Poor in Early Philadelphia* (Philadelphia: University of Pennsylvania Press, 2003), 80, 140–41.
62. Modern studies of alcohol consumption support the impulse to be strongly skeptical of these early nineteenth-century testimonials that portray the poor as inordinately intemperate. Researchers find that socioeconomic status is important in shaping consumption patterns, but the relationship resists easy generalizations. For instance, see Kellie E. M. Barr, Michael P. Farrell, Grace M. Barnes, and John W. Welte, "Race, Class, and Gender Differences in Substance Abuse: Evidence of Middle-Class/Underclass Polarization among Black Males," *Social Problems* 40, no. 3 (1993): 314–27.
63. *The Homeless Heir, or Life in Bedford Street: A Mystery of Philadelphia by John the Outcast* (Philadelphia. J. H. C. Whiting, 1856), 9–10.
64. Pennsylvania Society for Discouraging the Use of Ardent Spirits, *The Anniversary Report of the Managers of the Pennsylvania Society for Discouraging the Use of Ardent Spirits* (Philadelphia: Pennsylvania Society for Discouraging the Use of Ardent Spirits, 1831), 72.
65. All told, I found the deaths of 62 individuals recorded in the Philadelphia Cemetery Returns reported in the *Public Ledger*.
66. *Public Ledger*, December 4, 1840, 2.
67. Eliza Draper, Cemetery Returns HSP. *Public Ledger*, July 1, 1840, 2.
68. Dorsey, *Reforming Men and Women*, 61; Karin Wulf, "Gender and the Political Economy of Poor Relief in Colonial Philadelphia," in *Down and Out in Early America*, ed. Billy G. Smith (University Park: Pennsylvania State University Press, 2004).
69. By comparison, of the 1,541 individuals who died of either delirium tremens or intemperance, 432 were buried in the almshouse cemetery. Fewer than a third of those were women. Cemetery Returns HSP.
70. Mary Ann Ward, Cemetery Returns HSP; *Public Ledger*, December 27, 1849, 2.
71. Jane Colder, Cemetery Returns HSP; "Death from Intemperance," *Public Ledger*, July 6, 1843, 2.
72. Levi Lee, Cemetery Returns HSP; *Public Ledger*, December 18, 1847, 2.
73. "Cholera and Death," *Saturday Bulletin*, February 4, 1832, 2.
74. Charles E. Rosenberg, *The Cholera Years: The United States in 1832, 1849, and 1866* (Chicago: University of Chicago Press, 1962), 13–39.
75. "Cholera," *National Gazette and Literary Register*, June 30, 1832; "Quit Dram Drinking," broadside, Philadelphia, 1832, Library Company of Philadelphia; Admissions

to Cholera Hospital, Joseph Parrish, physician in chief, 1832, College of Physicians of Philadelphia; John Bell and David Francis Condie, *All the Material Facts in the History of Epidemic Cholera: Being a Report of the College of Physicians of Philadelphia, to the Board of Health* (Philadelphia: Thomas DeSilver, 1832), 35; John Bell, "Means of Prevention of Cholera," *Journal of Health* 3, no. 21 (1832): 337.

76. This gap in the cemetery returns records is likely the result of the health crisis.
77. See the weekly mortality statistics published in 1832 in the *National Gazette and Literary Register*.
78. *Saturday Courier*, March 3, 1838, May 12, 1838, May 19, 1838, and October 20, 1838
79. *Aurora and Franklin Gazette*, April 22, 1828.
80. Dr. Draper, "From Dr. Draper's Unpublished Medical Treatise on Mania a Potu," *Saturday Courier*, November 21 1835.
81. One measure of this difference is their presence in the city directories, which primarily listed proprietors and heads of household. The most inclusive directories printed between 1830 and 1850 included about 55 percent of the total male population. They excluded most of the large population of journeyman and wage laborers and almost all of the poor and transient. Almost a quarter of the individuals recorded in the cemetery returns as having died of delirium tremens can be found in the city directories. By contrast, only 8 percent of intemperance victims appear, reflecting this group's relative social obscurity. Stuart Blumin, "Mobility in a Nineteenth-Century American City: Philadelphia, 1820–1860" (PhD diss., University of Pennsylvania, 1968), 58.
82. Stuart Blumin, *The Emergence of the Middle Class: Social Experience in the American City, 1760–1900* (New York: Cambridge University Press, 1989), 109–12.
83. Michael Zakim, *Ready-Made Democracy: A History of Men's Dress in the American Republic, 1760–1860* (Chicago: University of Chicago Press, 2003), 47; Luskey, *On the Make*.
84. Blumin, *Emergence of the Middle Class*, 66–107.
85. David Hancock, *Oceans of Wine: Madeira and the Emergence of American Trade and Taste* (New Haven, CT: Yale University Press, 2009), 333–55; W. J. Rorabaugh, *The Alcoholic Republic: An American Tradition* (New York: Oxford University Press, 1979), 100–110.
86. Ian Tyrrell, *Sobering Up: From Temperance to Prohibition in Antebellum America* (Westport, CT: Greenwood Press, 1979), 3–15; Stott, *Jolly Fellows*, 65–96.
87. Of individuals who appeared in the Cemetery Returns and could also be found in the city directories, 22 percent listed explicitly nonmanual occupations.
88. Pliny Earle, "An Analysis of the Cases of Delirium Tremens Admitted into the Bloomingdale Asylum for the Insane from June 16th, 1821 to December 31st, 1844," *American Journal of the Medical Sciences* 15 (1848): 76–83.
89. Historian Stuart Blumin used census records to rank occupations in the Philadelphia city directories into categories by mean wealth for 1820-60. Because these rankings are calculated on mean wealth rather than type of work, each category contained a diverse range of occupations. To briefly summarize Blumin's overall categories: Category 1 includes elite professionals, such as doctors, as well as men in business, especially merchants. Category 2 includes small-scale retailers and businessmen, such as grocers and "conveyancers," and crafts that are skilled and also require a larger amount of capital, such as coach builders, cabinetmakers, and silversmiths. Category 3, by far the largest, represents Philadelphia's middling class of artisans:

coopers, brick makers, and stonecutters; butchers, confectioners, and bakers; and bookbinders, printers, and engravers. This group also includes tavern keepers, innkeepers, and victuallers. Category 4 includes low-paid nonmanual workers such as accountants and clerks, as well as carters, an unskilled job that required owning a cart and horse. Category 5 encompasses unskilled laborers, as well as weavers and other pieceworkers who worked in their homes.

90. For a complete list of the occupations of these 219 people ranked according to Blumin's wealth categories, see Matthew Warner Osborn, "The Anatomy of Intemperance: Alcohol and the Diseased Imagination in Philadelphia, 1784–1860" (PhD diss., University of California, 2007), 170.
91. Daniel Drake, *A Discourse on Intemperance: Delivered at Cincinnati, March 1, 1828* (Cincinnati: Looker and Reynolds, 1828), 26.
92. Register of Wills, administration file 51, book P, p. 44. Andrew Bossart, Cemetery Returns HSP.
93. Register of Wills, administration file 179, book N, p. 105. James Maher, Cemetery Returns HSP. Overall, the study found that the estates of this group of tavern keepers and innkeepers ranged in value between $3,750 and $150.
94. Ric Caric has also done considerable work in the almshouse records and reached some different conclusions. See Ric N. Caric, "Hideous Monsters before the Eye: Delirium Tremens and Manhood in Antebellum Philadelphia," in *Gender, Health, and Popular Culture*, ed. Cheryl Krasnick Warsh (Waterloo, ON: Wilfred Laurier University Press, 2011).
95. Tabb, "Statistics of the Causes of Death," 374.
96. Of these, 314 died of delirium tremens and 118 died of intemperance.
97. This study is done with the Almshouse Daily Occurrence Docket, which recorded every person admitted to the almshouse. The Docket includes a short interview at each inmate's first admission, but the information gathered was inconsistent. Fewer than half the entries list an occupation. Of the 1,532 deaths found in the cemetery returns sample, at least 232 appear in the Docket. These 232 represent approximately 50 percent and are only a portion of the total number of patients who died at the almshouse of alcohol abuse between 1825 and 1850.
98. Bruce Laurie, *Working People of Philadelphia, 1800–1850* (Philadelphia: Temple University Press, 1980), 9–10.
99. W. W. Gerhard, "Lectures on Clinical Medicine, Delivered at the Philadelphia Medical Institute," *Medical Examiner* 1, no. 13 (1838): 213.
100. Out of 223 records, 150 had marital status noted in the Daily Occurrence Docket. Philadelphia Almshouse Daily Occurrence Docket, 1825–1850, Guardians of the Poor, Philadelphia City Archives. Hereafter DOD.
101. DOD, May 2, 1828; December 5, 1827. Mary Craig, Cemetery Returns HSP.
102. DOD, November 12, 1828; November 15, 1828. Elizabeth Niket, Cemetery Returns HSP.
103. DOD, October 5, 1837; March 15, 1837; August 22, 1837; September 22, 1837; January 10, 1838; January 12, 1838. Jacob Miller, Cemetery Returns HSP.
104. Priscilla Ferguson Clement, *Welfare and the Poor in the Nineteenth-Century City: Philadelphia, 1800–1854* (Rutherford, NJ: Fairleigh Dickinson University Press, 1985), 83.
105. DOD, August 26, 1837; August 28, 1837. James Furlong, Cemetery Returns HSP.
106. DOD, November 14, 1839; January 1, 1840; June 8, 1841; June 16, 1841. Isaac Wheater, Cemetery Returns HSP.

107. The price of a gallon of whiskey in 1817 was about sixty cents. "Accounts of the Guardians of the Poor," *Aurora General Advertiser*, November 29, 1817, 3.
108. Gerhard, "Lectures on Clinical Medicine," 209–13; W. W. Gerhard, "Lectures on Clinical Medicine, Delivered at the Philadelphia Medical Institute: Delirium Tremens," *Medical Examiner* 1, no. 14 (1838): 223–24.
109. *Report of the Committee, Appointed by the Philadelphia Medical Society, January 24, 1829, to Take into Consideration the Propriety of That Society Expressing Their Opinion with Regard to the Use of Ardent Spirits*, 14–15.
110. DOD, January 26, 1841; June 3, 1841; September 30, 1841; March 25, 1841; April 21, 1842; April 20, 1842. Hester Parker, Cemetery Returns HSP.
111. DOD, November 14, 1828; February 6, 1829; February 9, 1829; February 28, 1829; April 20, 1829; April 24, 1829. Edward Maxwell, Cemetery Returns HSP.
112. Edward Pessen, "The Egalitarian Myth and the American Social Reality: Wealth, Mobility, and Equality in the 'Era of the Common Man,'" *American Historical Review* 76, no. 4 (1971): 989–1034.
113. Register of Wills, administration file 261, book O, p. 76. John Daley, Cemetery Returns HSP.
114. Register of Wills, administration file 354, book P, p. 307. William Keim, Cemetery Returns HSP.
115. Balleisen, *Navigating Failure*, 40. On a speculative note, this evidence suggests that one reason deaths from delirium tremens plummeted may be that alcohol consumption dropped because the extreme economic crisis left drinkers without cash to buy liquor.
116. Ibid., 3
117. George Gibson, "An Essay on Delirium Tremens" (MD diss., University of Pennsylvania, 1837), 1.
118. Patricia Okker, *Our Sister Editors: Sarah J. Hale and the Tradition of Nineteenth-Century American Women Editors* (Athens: University of Georgia Press, 1995).
119. Obituary, October 21, 1836, *Poulson's General Advertiser*. Francis Godey, Cemetery Returns HSP.
120. Patient Admittance Records, November 20, 1822, Pennsylvania Hospital Historical Collections.
121. Benjamin Coates, Practice Book, Coates Reynell Collection, Historical Society of Pennsylvania, Philadelphia.
122. DOD, June 30, 1836; August 2, 1836. Charles Nancrede, Cemetery Returns HSP. Joseph Nancrede, "On Mania a Potu," *American Medical Recorder* 1 (1818): 478–85; David A. Bloom, Gretchen Uznis, and Darrell A. Campbell, "Charles B. G. de Nancrede: Academic Surgeon at the Fin de Siècle," *World Journal of Surgery* 22 (1998): 1175–81.
123. Patient Admittance Records, April 16, 1847, Pennsylvania Hospital Historical Collections. Portrait of Dr. Joseph Klapp by Thomas Sully, 1814, Art Institute of Chicago. Klapp was the first American authority on delirium tremens. Sully's oldest son Thomas died of delirium tremens in 1847 at age thirty-five.
124. Register of Wills, administration file 153, book N, p. 253. Patrick Cain, Cemetery Returns HSP.
125. Hansell also owned a $50 gold watch, a $35 gun, and some fine furniture. Register of Wills, administration file 404, book P, p. 469. William Hansell, Cemetery Returns HSP.

126. Samuel B. Woodward, *Remarks on the Utility and Necessity of Asylums or Retreats for the Victims of Intemperance* (Philadelphia: Brown, Bicking and Guilbert, 1840), 3.
127. Patient Admission and Discharge, 1830–1850, Pennsylvania Hospital Historical Collections.
128. Gerhard, "Lectures on Clinical Medicine," 223.
129. For instance, Blanchard Fosgate, "Case of Delirium Tremens," *American Journal of the Medical Sciences* 7 (1844): 117–23; William H. Klapp, "Remarks on Delirium Tremens," *Medical Examiner* 2 (1839): 58–59; Jonathan Letherman, "Delirium Tremens Succesfully Treated by Chloroform," *Medical Examiner and Record of Medical Science* 17, no. 97 (1853): 33–37; Gerhard, "Lectures on Clinical Medicine," 223–24; "Medical Societies," 409–10.
130. [John Cotton Mather], *Autobiography of a Reformed Drunkard, or Letters and Recollections by an Inmate of the Alms-House* (Philadelphia: Griffin and Simon, 1845); William W. Pratt, *Ten Nights in a Bar-Room: A Drama in Five Acts, Dramatized from T. S. Arthur's Novel of the Same Name* (New York: Samuel French, 1860); Tyrrell, *Sobering Up*, 191–206.
131. Rosenberg, *Care of Strangers*, 276.
132. Woodward, *Remarks on the Utility and Necessity of Asylums or Retreats for the Victims of Intemperance*, 5.

CHAPTER FOUR

1. Benjamin H. Coates, "Address to the Kensington Young Men's Temperance Society, July 4, 1830," Coates Reynell Collection, Historical Society of Pennsylvania, Philadelphia.
2. Bruce Laurie, *Working People of Philadelphia, 1800–1850* (Philadelphia: Temple University Press, 1980), 40.
3. Ian Tyrrell, *Sobering Up: From Temperance to Prohibition in Antebellum America* (Westport, CT: Greenwood Press, 1979), 33–53.
4. Bruce Dorsey, *Reforming Men and Women: Gender in the Antebellum City* (Ithaca, NY: Cornell University Press, 2002), 115.
5. The best overall account of the temperance movement remains Tyrrell, *Sobering Up*.
6. Ibid., 87–124.
7. Lyman Beecher, *Six Sermons on the Nature, Occasions, Signs, Evils, and Remedy of Intemperance* (New York: American Tract Society, 1827).
8. Tyrrell, *Sobering Up*, 115.
9. Paul Johnson, *A Shopkeeper's Millennium: Society and Revivals in Rochester, 1815–1837* (New York: Hill and Wang, 1978); Laurie, *Working People of Philadelphia*, 39.
10. Tyrrell, *Sobering Up*, 115.
11. Dorsey, *Reforming Men and Women*, 116.
12. John Bell, *An Address to the Medical Students' Temperance Society of the University of Pennsylvania* (Philadelphia, 1833), 9; *The Anniversary Report of the Managers of the Pennsylvania Society for Discouraging the Use of Ardent Spirits* (Philadelphia: Pennsylvania Society for Discouraging the Use of Ardent Spirits, 1831), 68.
13. Robert Abzug, *Cosmos Crumbling: American Reform and the Religious Imagination* (New York: Oxford University Press, 1994), 8, 86–87, 94–95, 98, 101.
14. Beecher, *Six Sermons on Intemperance*, 13–14.
15. Laurie, *Working People of Philadelphia*, 40.
16. Abzug, *Cosmos Crumbling*, 94–95.

17. On the medical marketplace, see Lisa Rosner, "Thistle on the Delaware: Edinburgh Medical Education and Philadelphia Practice, 1800–1825," *Social History of Medicine* 5 (1992): 19–42.
18. Coates, "Address to the Kensington Young Men's Temperance Society, July 4, 1830."
19. "Medical and Philosophical Intelligence," *Philadelphia Journal of the Medical and Physical Sciences* 4 (1822): 204.
20. Ronald G. Walters, *American Reformers, 1815–1860* (New York: Hill and Wang, 1978), 158; Charles Colbert, *A Measure of Perfection: Phrenology and the Fine Arts in America* (Chapel Hill: University of North Carolina Press, 1997), 11; Benjamin H. Coates, "Lecture on Some of the Supports Derived by Phrenology from Comparative Anatomy" (1824?), Benjamin H. Coates Papers, Historical Society of Pennsylvania.
21. John Randolph, "A Memoir on the Life and Character of Philip Syng Physick," *American Journal of the Medical Sciences* 24 (1839): 102–22; "History of Phrenology in Philadelphia," *American Phrenological Journal* 2, no. 10 (1840): 476–77; Samuel Jackson to Benjamin Coates, May 3, 1827, Coates Correspondence, Coates Reynell Collection, Historical Society of Pennsylvania.
22. George Combe to Samuel Morton, May 2, 1839, Samuel Morton Correspondence, 1819–1850, American Philosophical Society, Philadelphia.
23. Benjamin Rush included a description of the science in his lectures on the mind in the first decade of the nineteenth century. He told his students the science "may be real, but [is] by no means certain." Benjamin Rush, "Lectures upon the Mind," 46, College of Physicians of Philadelphia.
24. See chapters 2 and 3 for my discussion of the Panic of 1819 and its influence on changing perceptions of poverty and alcohol abuse.
25. John Bell, "On Phrenology, or The Study of the Intellectual and Moral Nature of Man," *Philadelphia Journal of the Medical and Physical Sciences* 4, no. 7 (1822): 77.
26. Ibid., 204.
27. Bell, "On Phrenology," 114.
28. Benjamin H. Coates, "An Essay on Ideality, and the Poetical Temperament, to Be Read before the Phrenological Society, as a Lecture; 3rd Mo.: 11th, 1824," Medical Essays, Benjamin H. Coates Papers, Coates Reynell Collection, Historical Society of Pennsylvania.
29. F. J. V. Broussais, *A Treatise on Physiology Applied to Pathology*, trans. John Bell and R. LaRoche (Philadelphia: H. C. Carey and I. Lea, 1826); Stephen Nissenbaum, *Sex, Diet, and Debility in Jacksonian America: Sylvester Graham and Health Reform* (Chicago: Dorsey Press, 1980), 75.
30. William E. Horner, *A Treatise on Pathological Anatomy* (Philadelphia: Carey, Lea and Carey, 1829); "Introductory to a Course of Physiology, and Some Parts of Pathology and Practice of Medicine," Benjamin Coates Medical Essays, Coates Reynell Collection, Historical Society of Pennsylvania.
31. These developments are described in chapter 2.
32. John Harley Warner, "Remembering Paris: Memory and the American Disciples of French Medicine in the Nineteenth Century," *Bulletin of the History of Medicine* 65, no. 3 (1991): 315; Karen Halttunen, *Confidence Men and Painted Women: A Study of Middle-Class Culture in America, 1830–1870* (New Haven, CT: Yale University Press, 1982).
33. Reynell Coates, Bankruptcy Filing, 1844, National Archives and Records Administration, Mid-Atlantic Region, Philadelphia.
34. Steven J. Peitzman, "'I Am Their Physician': Dr. Owen J. Wister of Germantown

and His Too Many Patients," *Bulletin of the History of Medicine* 83, no. 2 (2009): 245–70.

35. William Barlow and David O. Powell, "A Dedicated Medical Student: Solomon Mordecai, 1819–1822," *Journal of the Early Republic* 7, no. 4 (1987): 383.
36. D. F. Condie to Benjamin Coates, January 1, 1827, Practice Books, box 3, Coates Reynell Collection, Historical Society of Pennsylvania; "Medical Instruction," *North American Medical and Surgical Journal* 9 (1830): 239.
37. "Medical Institute of Philadelphia," *Aurora General Advertiser*, April 9, 1827, 3.
38. "Address Delivered at Opening the Winter Course of the Medical Lyceum of Philadelphia for 1825–1826," Benjamin Coates Medical Essays, Coates Reynell Collection, Historical Society of Pennsylvania.
39. The largest and most prestigious medical journal of the 1820s was the *Philadelphia Journal of the Medical and Physical Sciences* (1820–27), which then became the *American Journal of the Medical Sciences* (1827–1924). The *American Medical Recorder* (1818–29), the *American Medical Review* (1825–26), the *Journal of Foreign Medical Science and Literature* (1821–24), and the *Medical Review and Analectic Journal* (1824–25) were also published in Philadelphia.
40. Samuel Jackson of Northumberland to Benjamin H. Coates, November 11, 1827, Benjamin H. Coates Correspondence, Coates Reynell Collection, Historical Society of Pennsylvania.
41. Samuel Jackson of Northumberland to Benjamin H. Coates, May 3, 1827, Benjamin H. Coates Correspondence, Coates Reynell Collection, Historical Society of Pennsylvania.
42. Benjamin H. Coates, "Observations on Delirium Tremens, or the Disease Improperly Called Mania a Potu," *North American Medical and Surgical Journal* 4, no. 7 (1827): 27–52, 205–34.
43. William H. Klapp, "Remarks on Delirium Tremens," *Medical Examiner* 2 (1839): 58–59.
44. Judah Dobson to Benjamin H. Coates, November 23, 1827, and James Kay Jr. to Benjamin H. Coates, January 15, 1830, Benjamin H. Coates Correspondence, Coates Reynell Collection, Historical Society of Pennsylvania.
45. Entry dated September 6, 1834, Archives of the Pennsylvania Hospital, Library Register, 1824—1842, American Philosophical Society, Philadelphia.
46. Samuel Coates, Memoranda Book, "Cases of Several Lunatics in the Pennsylvania Hospital, and the Causes There of in Many of the Cases," 70, 80–85, 104–8, 117–27, Archives of the Pennsylvania Hospital, American Philosophical Society, Philadelphia.
47. Richard Nesbitt, Poem and Painting, Rush Papers, Historical Society of Pennsylvania, Rush MSS, box 8.
48. Thomas G. Morton, *The History of the Pennsylvania Hospital, 1751–1895* (Philadelphia: Times, 1895), 139.
49. *Report of a Committee Appointed by the Pennsylvania Society for Discouraging the Use of Ardent Spirits, to Examine and Report What Amendments Ought to Be Made in the Laws of the Said State for the Suppression of Vice and Immorality, Particularly Those against Gaming* (Philadelphia: Atkinson and Alexander, 1828).
50. *Aurora General Advertiser*, July 4, 1828.
51. Nissenbaum, *Sex, Diet, and Debility*, 74.
52. Samuel Emlem, "Remarks upon the Mischievous Effects on Society of Spirituous

Liquors and the Means of Preventing Them," *North American Medical and Surgical Journal* 3 (1827): 267–77.

53. *An Address to Physicians by the Executive Committee of the Board of Managers of the New York City Temperance Society* (New York: J. and J. Harper, 1829).
54. *Report of the Pennsylvania Society for Discouraging the Use of Ardent Spirits* (Philadelphia: Pennsylvania Society for Discouraging the Use of Ardent Spirits, 1829), 1–2.
55. Dorsey, *Reforming Men and Women*, 113.
56. "Temperance," *Aurora and Franklin Gazette*, July 4, 1828.
57. Dorsey, *Reforming Men and Women*, 115.
58. *Report of the Committee, Appointed by the Philadelphia Medical Society, January 24, 1829, to Take into Consideration the Propriety of That Society Expressing Their Opinion with regard to the Use of Ardent Spirits* (Philadelphia: John Clarke, 1829), 8–9; Pennsylvania Society for the Promotion of Public Economy, *Report of the Library Committee of the Pennsylvania Society, for the Promotion of Public Economy, Containing a Summary of the Information Communicated by Sundry Citizens* (Philadelphia: Pennsylvania Society for the Promotion of Public Economy, 1817).
59. *Report of the Committee, Appointed by the Philadelphia Medical Society, January 24, 1829, to Take into Consideration the Propriety of That Society Expressing Their Opinion with regard to the Use of Ardent Spirits*, 14.
60. *Address to Physicians by the Executive Committee of the Board of Managers of the New York City Temperance Society.*
61. Dorsey, *Reforming Men and Women*, 115.
62. David Francis Condie, *Preamble and Resolutions* (Philadelphia: Temperance Beneficial Association, Southwark Branch, 1837).
63. Bell, *Address to the Medical Students' Temperance Society of the University of Pennsylvania*, 4.
64. John Clark to Benjamin H. Coates, October 27, 1829, Benjamin H. Coates Correspondence, Coates Reynell Collection, Historical Society of Pennsylvania.
65. "Ardent Spirits," *Friend: A Religious and Literary Journal* 3, no. 9 (1829): 67–68; "Report of the Committee Appointed by the Philadelphia Medical Society," *Register of Pennsylvania* 4, no. 22 (1829): 337–44; "On the Use and Abuse of Ardent Spirits," *Journal of Health* 1, no. 10 (1830): 157–61. John Clark to Benjamin H. Coates, October 27, 1829, Benjamin H. Coates Correspondence, Coates Reynell Collection, Historical Society of Pennsylvania.
66. *Report of the Committee, Appointed by the Philadelphia Medical Society, January 24, 1829, to Take into Consideration the Propriety of That Society Expressing Their Opinion with regard to the Use of Ardent Spirits*, 16.
67. Ibid., 3.
68. Ibid., 6.
69. Ibid., 14.
70. Benjamin Rush, "Of Drunkenness and Its Cures," Benjamin Rush Papers, Library Company of Philadelphia; Benjamin Rush, *Letters of Benjamin Rush: 1793–1813*, ed. L. H. Butterfield, vol. 2 (Princeton, NJ: Princeton University Press, 1951), 976–79.
71. Johnson, *Shopkeeper's Millennium*, 120; Laurie, *Working People of Philadelphia*, 71–72.
72. *Mechanic's Free Press*, January 9, 1830, 2.
73. "Masters and Apprentices," *Mechanic's Free Press*, November 29, 1828, 2.
74. "The Drunkard," *Mechanic's Free Press*, July 19, 1828, 2.
75. "Drunkenness," *Mechanic's Free Press*, October 25, 1828, 1.
76. "Temperance," *Mechanic's Free Press*, December 28, 1828, 2.

77. Albert Barnes, *The Connexion of Temperance with Republican Freedom: An Oration* (Philadelphia: Mechanics and Workingmen's Temperance Society of the City and County of Philadelphia, 1835); Dorsey, *Reforming Men and Women*, 120–24; Laurie, *Working People of Philadelphia*, 119–24.
78. "Workingmen, Attend," *Public Ledger*, September 26, 1837, 2.
79. Laurie, *Working People of Philadelphia*, 119–24.
80. Ibid., 122.
81. Dorsey, *Reforming Men and Women*, 206–12.
82. Ruth Alexander, "We Are Engaged as a Band of Sisters: Class and Domesticity in the Washingtonian Temperance Movement, 1840–1850," *Journal of American History* 75 (1988): 763–85.
83. Abzug, *Cosmos Crumbling*, 103–4; Tyrrell, *Sobering Up*, 159–224.
84. Mark Edward Lender and James Kirby Martin, *Drinking in America: A History* (New York: Free Press, 1982), 75.
85. "Temperance Conversation Meeting," *Public Ledger*, March 13, 1841, 2.
86. "Temperance in Philadelphia," *Public Ledger*, March 19, 1841, 2.
87. "Southwark Branch no. 1, Temperance Beneficial Association," *Public Ledger*, February 9, 1841, 2; Condie, *Preamble and Resolutions*.
88. In the early 1840s, activity was so high that on any given night some branch of the Temperance Beneficial Association was likely holding a meeting. See, for instance, *Public Ledger*, February 9–13, 1841.
89. Condie, *Preamble and Resolutions; First Annual Report of the Board of Managers of the Philadelphia Temperance and Benevolent Association* (Philadelphia: Merrihew and Thompson, 1841).
90. *First Annual Report of the Board of Managers of the Philadelphia Temperance and Benevolent Association*, 8.
91. Peitzman, "'I Am Their Physician'"; Barlow and Powell, "Dedicated Medical Student."
92. Steven M. Stowe, *Doctoring the South: Southern Physicians and Everyday Medicine in the Mid-Nineteenth Century* (Chapel Hill: University of North Carolina Press, 2004), 108–13; Paul Starr, *The Social Transformation of American Medicine* (New York: Basic Books, 1982), 88–92.
93. Reynell Coates to Benjamin H. Coates, May 9, 1830, Benjamin Coates Correspondence, Coates Reynell Collection, Historical Society of Pennsylvania.
94. Reynell Coates, Bankruptcy Filing, 1844, National Archives and Records Administration, Mid-Atlantic Region, Philadelphia.
95. Charles E. Rosenberg, "John Gunn: Everyman's Physician," in *Explaining Epidemics and Other Studies in the History of Medicine* (New York: Cambridge University Press, 1992).
96. Charles E. Rosenberg, "Medical Text and Social Context: Explaining William Buchan's *Domestic Medicine*," in *Explaining Epidemics and Other Studies in the History of Medicine* (New York: Cambridge University Press, 1992); Charles E. Rosenberg, "The Book in the Sickroom: A Tradition of Print and Practice," Library Company of Philadelphia, http://www.librarycompany.org/doctor/rosen.html.
97. Charles E. Rosenberg and William H. Helfand, "'Every Man His Own Doctor': Popular Medicine in Early America," Library Company of Philadelphia, http://www.librarycompany.org/doctor/intro.html.
98. Starr, *Social Transformation of American Medicine*, 60–64.
99. Ibid., 51–54.

100. Warner, "Remembering Paris," 315.
101. John Harley Warner, "Medical Sectarianism, Therapeutic Conflict, and the Shaping of Orthodox Professional Identity in Antebellum American Medicine," in *Medical Fringe and Medical Orthodoxy, 1750–1850*, ed. William F. Bynum and Roy Porter (London: Croom Helm, 1987), 253.
102. Dr. Eugene Palmer to Benjamin H. Coates, October 22, 1832, Benjamin Coates Correspondence, Coates Reynell Collection, Historical Society of Pennsylvania. On the subject of trust, also see Peitzman, "'I Am Their Physician.'"
103. John Harley Warner, *The Therapeutic Perspective: Medical Practice, Knowledge, and Identity in America, 1820–1885* (Cambridge, MA: Harvard University Press, 1986), 1–10.
104. Rebecca J. Tannenbaum, "Earnestness, Temperance, Industry: The Definition and Uses of Professional Character among Nineteenth-Century American Physicians," *Journal of the History of Medicine and Allied Sciences* 49 (1994): 251–83.
105. "Report of the Committee Appointed by the Philadelphia Medical Society," 13.
106. Stowe, *Doctoring the South*, 90–98.
107. Daniel Drake, *A Discourse on Intemperance: Delivered at Cincinnati, March 1, 1828* (Cincinnati: Looker and Reynolds, 1828), 26–27.
108. "The Proceedings of the Physiological Temperance Society of the Medical Institute of Louisville," *Western Journal of Medicine and Surgery* 5, no. 2 (1842): 270.
109. John Marsh, *Temperance Recollections: Labors, Defeats, Triumphs: An Autobiography* (New York: Charles Scribner, 1866), 111.
110. Bell, *Address to the Medical Students' Temperance Society of the University of Pennsylvania*, 8–9, 10.
111. Stowe, *Doctoring the South*, 93.
112. Bell, *Address to the Medical Students' Temperance Society of the University of Pennsylvania*, 9.
113. "Prospectus," *Journal of Health* 1, no. 1 (1829): 1.
114. See, for instance, *Saturday Bulletin*, August 7, 1830, March 16, 1831, January 14, 1832; *Aurora and Pennsylvania Gazette*, November 3, 1829.
115. "The Journal of Health," *Mechanic's Free Press*, February 20, 1830.
116. "The Middle Classes," *Journal of Health* 1, no. 23 (1830): 357–58.
117. Nissenbaum, *Sex, Diet, and Debility*, 140–54.
118. Ibid., 14.
119. Michael Sappol, *A Traffic of Dead Bodies: Anatomy and Embodied Social Identity in Nineteenth-Century America* (Princeton, NJ: Princeton University Press, 2002), 173–75.
120. "Introductory Explanations to the New Series," *Phrenological Journal and Magazine of Moral Science* 1 (1838): 4–5.
121. Madeleine B. Stern, *Heads and Headlines: The Phrenological Fowlers* (Norman: University of Oklahoma Press, 1971), 15–33.
122. "Progress of Phrenology," *Boston Medical and Surgical Journal* 22, no. 22 (1840): 353.
123. Colbert, *Measure of Perfection*, 20–27.
124. "Phrenology and Temperance," *Boston Medical and Surgical Journal* 23, no. 6 (1841): 419.
125. Nissenbaum, *Sex, Diet, and Debility*, 149–52.
126. Ibid., 150.
127. Starr, *Social Transformation of American Medicine*, 79–144.

CHAPTER FIVE

1. Quoted in Keith L. Sprunger, "Cold Water Congressmen: The Congressional Temperance Society before the Civil War," *Historian* 27, no. 4 (1965): 513.
2. "Correspondence and Discussion between Dr. T. Hun and E. C. Delavan relative to Dr. Sewall's Drawings of the Human Stomach and the Doctrine They Teach," *Enquirer* 1, no. 3 (1843): 99.
3. Thomas Sewall, *The Pathology of Drunkenness, or The Physical Effects of Alcoholic Drinks: With Drawings of the Drunkard's Stomach* (Albany, NY: C. Van Benthuysen, 1841); William E. Horner, *A Treatise on Pathological Anatomy* (Philadelphia: Carey, Lea and Carey, 1829), 454–55.
4. "Correspondence and Discussion between Dr. T. Hun and E. C. Delavan, 99.
5. Ibid.
6. Sprunger, "Cold Water Congressman," 513–15.
7. Ibid.; John Marshall, *Temperance Recollections: Labors, Defeats, Triumphs; An Autobiography* (New York: Charles Scribner, 1866), 88–92.
8. "Congressional Temperance Society," *New York Times*, February 24, 1879.
9. *Report of the Committee, Appointed by the Philadelphia Medical Society, January 24, 1829, to Take into Consideration the Propriety of That Society Expressing Their Opinion with regard to the Use of Ardent Spirits* (Philadelphia: John Clarke, 1829), 5.
10. Sarah Tracy, *Alcoholism in America: From Reconstruction to Prohibition* (Baltimore: Johns Hopkins University Press, 2005).
11. Charles E. Rosenberg, "What Is Disease? In Memory of Owsei Temkin," *Bulletin of the History of Medicine* 77, no. 3 (2003): 491–505.
12. Wilkie A. Wilson and Cynthia M. Kuhn, "How Addiction Hijacks Our Reward System," *Cerebrum: The Dana Forum on Brain Science* 7, no. 2 (2005): 53–56.
13. Gilbert Flagler, "A Case of Mania," *American Medical Recorder* 2 (1819): 185–86.
14. It is impossible to know how many patients may have developed delirium tremens as a secondary affliction, which doctors at the hospital reported to be common. These numbers, then, refer only to patients listed in the register with delirium tremens as their primary complaint. Pennsylvania Hospital, Patient Admittance Register, 1820–50.
15. Benjamin H. Coates, "Observations on Delirium Tremens, or the Disease Improperly Called Mania a Potu," *North American Medical and Surgical Journal* 4, no. 8 (1827): 214.
16. *The Anniversary Report of the Managers of the Pennsylvania Society for Discouraging the Use of Ardent Spirits*, (Philadelphia: Pennsylvania Society for Discouraging the Use of Ardent Spirits, 1831), 28.
17. William Stokes, "On Delirium Tremens," *American Journal of the Medical Sciences* 14 (1834): 495.
18. William Sweetser, *A Dissertation on Intemperance: To Which Was Awarded the Premium Offered by the Massachusetts Medical Society*, ed. Gerald Grob, Nineteenth-Century Medical Attitudes toward Alcoholic Addiction (New York: Arno Press, 1981), 90.
19. Coates, "Observations on Delirium Tremens," 50; "Delirium Tremens," *Medical Recorder of Original Papers and Intelligence in Medicine and Surgery* 12, no. 2 (1827): 421–22.
20. Alexander Henry, *An Address to the Citizens of Philadelphia on the Subject of Establishing an Asylum for the Cure of Victims of Intemperance* (Philadelphia: Brown, Bicking and Guilbert, 1841), 6.
21. John Allen Krout, *The Origins of Prohibition* (New York: Knopf, 1925), 140.

22. *Report of the Committee, Appointed by the Philadelphia Medical Society, January 24, 1829, to Take into Consideration the Propriety of That Society Expressing Their Opinion with regard to the Use of Ardent Spirits*, 13.
23. For more on medical concerns about female intemperance, see Scott C. Martin, *Devil of the Domestic Sphere: Temperance, Gender, and Middle-Class Ideology, 1800–1860* (DeKalb: Northern Illinois University Press, 2008), 68–86.
24. "Medical Societies," *Medical and Surgical Reporter* 2, no. 3 (1859): 57–61.
25. "Report of the Committee Appointed by the Philadelphia Medical Society," *Register of Pennsylvania* 4, no. 22 (1829): 8.
26. Quoted in Martin, *Devil of the Domestic Sphere*, 73.
27. Reuben D. Mussey, *An Address on Ardent Spirit, Read before the New-Hampshire Medical Society at Their Annual Meeting, June 5, 1827* (Boston: Perkins and Marvin, 1829); Samuel Emlem, "Remarks upon the Mischievous Effects on Society of Spirituous Liquors and the Means of Preventing Them," *North American Medical and Surgical Journal* 3 (1827): 267–77; Thomas Sewall, *An Address Delivered before the Washington City Temperance Society* (Washington, DC: Washington City Temperance Society, 1830); Sweetser, *Dissertation on Intemperance*; Daniel Drake, *A Discourse on Intemperance: Delivered at Cincinnati, March 1, 1828* (Cincinnati: Looker and Reynolds, 1828); Reuben D. Mussey, *Prize Essay on Ardent Spirits, and Its Substitutes as a Means of Invigorating Health* (Washington, DC: D. Green, 1837).
28. *Report of the Committee, Appointed by the Philadelphia Medical Society, January 24, 1829, to Take into Consideration the Propriety of That Society Expressing Their Opinion with regard to the Use of Ardent Spirits*, 14.
29. Mussey, *Address on Ardent Spirit*, 4–5.
30. Ibid., 8–9.
31. Drake, *Discourse on Intemperance*.
32. Jessica Warner, "Old and in the Way: Widows, Witches, and Spontaneous Combustion in the Age of Reason," *Contemporary Drug Problems* 23 (1996): 197–220.
33. Jessica Warner, *Craze: Gin and Debauchery in an Age of Reason* (New York: Four Walls Eight Windows, 2002).
34. Pierre-Aimé Lair, "On the Combustion of the Human Body, Produced by the Long and Immoderate Use of Spirituous Liquors," *Emporium of Arts and Sciences* 1 (1812): 161–78.
35. W. Tooke, "Observations on Spontaneous Inflammations," *Emporium of Arts and Sciences* 1 (1812): 94–105.
36. *Aurora General Advertiser*, October 3, 1811, 2.
37. Thomas Trotter, *An Essay Medical, Philosophical, and Chemical on Drunkenness, and Its Effects on the Human Body*, ed. Roy Porter, Tavistock Classics in the History of Psychiatry (New York: Routledge, 1988), 63–91; Robert Macnish, *The Anatomy of Drunkenness*, 4th ed. (Glasgow: W. R. McPhun, 1832), 178.
38. McPhun, *Anatomy of Drunkenness*, 78–79.
39. James Overton, "Case of Spontaneous Combustion," *American Journal of the Medical Sciences* 17 (1835): 266–68.
40. *Report of the Committee, Appointed by the Philadelphia Medical Society, January 24, 1829, to Take into Consideration the Propriety of That Society Expressing Their Opinion with regard to the Use of Ardent Spirits*, 5.
41. *Saturday Bulletin*, July 17, 1830; *Saturday Evening Post*, July 24, 1830; "Spontaneous Combustion of a Drunkard," *Journal of Health* 1, no. 18 (1830): 288.

42. Drake, *Discourse on Intemperance*, 46.
43. Joseph Alison, *The Rum Maniac* (New York: American Temperance Union, [1851–70]).
44. Edward C. Delavan, *Defence of Dr. Sewall's Work on the Pathology of Drunkenness, and His Drawings of the Human Stomach, as Affected by the Use of Alcoholic Drinks from Health to Death by Delirium Tremens* (Albany, NY, 1843), 12–13.
45. Sewall, *Pathology of Drunkenness*, 1; National Library of Medicine, National Institutes of Health.
46. Marsh, *Temperance Recollections*, 115.
47. "Correspondence and Discussion between Dr. T. Hun and E. C. Delavan," 101.
48. Ibid., 104.
49. Ibid., 101.
50. "Lectures on India, with Magic Lantern," *Temperance Annual and Cold Water Magazine* 1, no. 1 (1843): 32.
51. Emlem, "Remarks upon the Mischievous Effects on Society of Spirituous Liquors and the Means of Preventing Them," 275–76.
52. *Aurora General Advertiser*, September 7, 1827, 1.
53. "Medical Department," *Christian Advocate and Journal* 2 (1835): 35; John H. Kain, "On *Intemperance* Considered as a Disease and Susceptible of Cure," *American Journal of the Medical Sciences* 3 (1828): 291–95; "Cure for Drunkenness," *Boston Recorder and Religious Telegraph*, March 2, 1827, 36.
54. *Aurora General Advertiser*, December 12, 1827, 1.
55. William E. Channing, *A Discourse on the Life of Rev. Joseph Tuckerman, D.D.* (Boston: William Crosby, 1841).
56. Joseph Tuckerman, Diary, July 3, 6, 1827, Joseph Tuckerman Papers, MS N-1682, Massachusetts Historical Society. I owe Trisha Posey a debt of gratitude for providing me with this anecdote.
57. "Report of the Medical Society of the City of New-York on Nostrums, or Secret Medicines," *New York Medical and Physical Journal* 6 (1827): 428–32.
58. Benjamin Rush, "Of Drunkenness and Its Cures," 11–12, Lectures on Fevers, Benjamin Rush Collection, Library Company of Philadelphia.
59. It's unclear if the physician knew of Chambers's Remedy. The anecdote was written in the context of correspondence regarding Klapp's cure. Samuel Jackson of Northumberland to Benjamin Coates, May 3, 1827, Benjamin Coates Correspondence, Coates Reynell Collection, Historical Society of Pennsylvania, Philadelphia.
60. "Report of the Medical Society of the City of New-York on Nostrums, or Secret Medicines."
61. "Chambers' Medicine," *New England Farm and Horticultural Journal* 6, no. 25 (1828): 197.
62. W. D. Brincklé, "Observations on the Use of Sulphuric Acid in the Cure of Intemperance: With Cases," *North American Medical and Surgical Journal* 4 (1827): 293.
63. Sweetser, *Dissertation*, 93.
64. *Report of the Committee, Appointed by the Philadelphia Medical Society, January 24, 1829, to Take into Consideration the Propriety of That Society Expressing Their Opinion with regard to the Use of Ardent Spirits*, 11–12.
65. Ibid., 30–33.
66. "Cure for Intemperance," *New England Farm and Horticultural Journal* 6, no. 31 (1828): 248.

67. Benoit Denizet-Lewis, "An Anti-addiction Pill?" *New York Times*, June 25, 2006.
68. Nancy Tomes, *A Generous Confidence: Thomas Story Kirkbride and the Art of Asylum-Keeping, 1840–1883* (New York: Cambridge University Press, 1984), 106.
69. Lawrence Goodheart, *Mad Yankees: The Hartford Retreat for the Insane and Nineteenth-Century Psychiatry* (Amherst: University of Massachusetts Press, 2003).
70. Tomes, *Generous Confidence*, 4–7; David J. Rothman, *The Discovery of the Asylum: Social Order and Disorder in the New Republic* (Boston: Little, Brown, 1971).
71. Samuel B. Woodward, *Essays on Asylums for Inebriates* (Worcester, MA, 1838), 2.
72. Henry, *Address to the Citizens of Philadelphia*, 3–5.
73. *Remarks on the Utility and Necessity of Asylums or Retreats for the Victims of Intemperance* (Philadelphia: Brown, Bicking and Guilbert, 1840), 7.
74. Tomes, *Generous Confidence*, 106; Tracy, *Alcoholism in America*, 96.
75. Thomas S. Kirkbride, *Reports of the Pennsylvania Hospital for the Insane* (Philadelphia: Board of Managers of the Pennsylvania Hospital, 1846), 39, 47.
76. Thomas Kirkbride, *Report of the Pennsylvania Hospital for the Insane for the Year 1847* (Philadelphia: Board of Managers, 1848), 16.
77. Ibid.
78. Ibid.
79. Ibid., 38.
80. Ibid.
81. "Hall, J. C., to Kirkbride," April 7, 1846, TSK Files, General Correspondence, Pennsylvania Hospital Historical Archives, Philadelphia.
82. "Medical Societies."
83. Ibid., 83.
84. Tomes, *Generous Confidence*, 239.
85. "Medical Societies," 83.
86. "George Biddle to Kirkbride," January 30, 1872, TSK Files, General Correspondence, Pennsylvania Hospital Archives, Philadelphia.
87. Tracy, *Alcoholism in America*, 2.
88. Abraham Lincoln, "February 22, 1842—Address before the Springfield Washingtonian Temperance Society," in *Abraham Lincoln: Complete Works, Comprising His Speeches, Letters, State Papers, and Miscellaneous Writings*, ed. John G. Nicolay and John Hay (New York: Century, 1920), 60.
89. Ronald G. Walters, *American Reformers, 1815–1860* (New York: Hill and Wang, 1978), 130–34.
90. Benjamin Estes, *Essay on the Washingtonian Temperance Movement* (New York, 1846), quoted in Thomas Augst, "Temperance, Mass Culture, and the Romance of Experience," *American Literary History* 19 (2007): 4–5.
91. Tracy, *Alcoholism in America*, 98–100.
92. David Reynolds, "Black Cats and Delirium Tremens: Temperance and the American Renaissance," in *The Serpent in the Cup: Temperance in American Literature*, ed. David Reynolds and Debra Rosenthal (Amherst: University of Masschusetts Press, 1997), 28.
93. T. S. Arthur, *Six Nights with the Washingtonians: A Series of Temperance Tales* (Philadelphia: Godey and McMichael, 1843), 48–49.
94. John B. Gough, *An Autobiography* (Boston: John B. Gough, 1845), 44–45.
95. *Confessions of a Reformed Inebriate* (New York: American Temperance Union, 1848), 139.
96. James Root, *The Horrors of Delirium Tremens* (New York: J. Adams, 1844); James Root, *Jefferson Brick versus Delirium Tremens* (New York: Delirium Tremens, 1844).
97. Root, *Horrors of Delirium Tremens*.

CHAPTER SIX

1. Joseph Allison, *The Rum Maniac* (New York: American Temperance Union, [1851–70]).
2. William T. Taylor, "Castration: Recovery, Followed by Phthisis Pulmonalis," *American Journal of the Medical Sciences* 30 (1855): 85–86.
3. Michael Sappol, "The Odd Case of Charles Knowlton: Anatomical Performance, Medical Narrative, and Identity in Antebellum America," *Bulletin of the History of Medicine* 83, no. 3 (2009): 460–98.
4. For more on economic self-possession and manhood in antebellum popular literature, see David Anthony, *Paper Money Men: Commerce, Manhood, and the Sensational Public Sphere in Antebellum America* (Columbus: Ohio State University Press, 2009).
5. See George Cheyne, *An Essay of Health and Long Life*, 4th ed. (Bath, 1725), 45; Benjamin Rush, An *Inquiry into the Effects of Ardent Spirits upon the Human Body and Mind, with an Account of the Means of Preventing, and of the Remedies for Curing Them*, ed. Gerald Grob, 8th ed., Addiction in America: Drug Abuse and Alcoholism (New York: Arno Press, 1981), 43; Samuel Jackson, "Observations on Delirium Tremens," *American Journal of Medical Science* 7 (1830–31): 363.
6. Thomas H. Wright, "Observations on the Treatment of Delirium Tremens, and on the Use of the Warm Bath in That Disease," *American Journal of the Medical Sciences* 6 (1830): 23; John Ware, "Remarks on the History and Treatment of Delirium Tremens by John Ware, M.D., Fellow of the Massachusetts Medical Society," *American Journal of the Medical Sciences* 9 (1832): 165.
7. Wright, "Observations on Delirium Tremens," 17.
8. Benjamin H. Coates, "Observations on Delirium Tremens, or the Disease Improperly Called Mania a Potu," *North American Medical and Surgical Journal* 4, no. 8 (1827). 214.
9. Ibid.
10. Myra Glen, "Troubled Manhood in the Early Republic: The Life and Autobiography of Horace Lane," *Journal of the Early Republic* 26 (2006): 59–94.
11. Horace Lane, *The Wandering Boy, Careless Sailor, and Result of Inconsideration* (Skaneateles, NY: Luther A. Pratt, 1839), 181–82.
12. George Lippard, *The Killers: A Narrative of Real Life in Philadelphia, by a Member of the Philadelphia Bar* (Philadelphia: Hankinson and Bartholomew, 1849), 34. On the slang term "man with the poker," see Ric N. Caric, "The Man with the Poker Enters the Room: Delirium Tremens and Popular Culture in Philadelphia, 1828–1850," *Pennsylvania Magazine* 74, no. 4 (2007): 452–91.
13. George Lippard, "Memoirs of a Preacher," *Quaker City News*, January 27, 1849, 1.
14. The exact derivation of the "red hot poker" is unknown. In an age when coal or wood fires were the sole source of heat in homes, pokers were certainly in abundance. Hot pokers were also used to warm alcoholic drinks.
15. Michael Scott, *Tom Cringle's Log*, 2 vols. (Philadelphia: E. L. Carey and A. Hart, 1833), 1:199.
16. Ibid., 2:6. Library Company of Philadelphia librarian James Green believes the note was most likely written about 1840 by Charles A. Poulson, son of the publisher of *Poulson's Daily Advertiser*. Poulson made extensive use of the library and wrote in a number of books. Attribution was based on the handwriting as well as on Poulson's strong interest in temperance.
17. Given the time it was published and its enormous popularity, the poem might have even been an inspiration for the first description of delirium tremens. Three British

physicians first published essays on delirium tremens in 1813, and two of them, John Armstrong and Samuel Pearson, were Scottish. Armstrong credited Pearson with first describing the disease in a short pamphlet published in 1801, not long after the publication of *Tam o' Shanter*. Pearson's case histories include a patient suffering from "brain fever," as he called it, who in his delusions insisted on visiting a graveyard.

18. John J. Davis, "An Essay on Delirium Tremens" (MD diss., University of Pennsylvania, 1861), 4–6.
19. *Cantraip sleight* = black magic; *gibbet-airns* = gibbet irons; *Twa span-lang, wee, unchristened bairns* = two tiny babies; *new-cutted frae a rape* = just cut down from a hangman's noose; *gab* = mouth; stack to the heft = stuck to the haft. Ibid., 4–5; Robert Burns, *Understanding Robert Burns: Verse, Explanation, and Glossary*, ed. George Scott Wilkie (Glasgow: Neil Wilson, 2002), 244.
20. Terry Castle, *The Female Thermometer: Eighteenth-Century Culture and the Invention of the Uncanny* (New York: Oxford University Press, 1995), 174.
21. Philadelphia Area Theatre Playbills, vol. 24, Masonic Hall, November 1834, Library Company of Philadelphia.
22. Charles Durang, "The Philadelphia Stage: From the Year 1749 to 1855; Third Series, Embracing the Period between the Season of 1830–31 and the Demolition of the Chestnut Street Theatre, April 1855," 312, 351. Article preserved in a scrapbook compiled by Charles Poulson, Library Company of Philadelphia.
23. "Edgar Poe," *Littell's Living Age* 5 (1854): 171.
24. Rush, *Inquiry*, 6.
25. Literary scholars have variously read "Metzengerstein" as a "surprisingly cogent" anti-abolitionist and racist argument, and, alternatively, as a story that stages the "triumph of the subaltern's point of view." See Maurice S. Lee, "Absolute Poe: His System of Transcendental Racism," *American Literature* 75, no. 4 (2003): 751–81; Agniezska M. Soltysik, *Poetics and Politics of the American Gothic: Gender and Slavery in Nineteenth-Century American Literature* (Farnham, Surrey, UK: Ashgate, 2010), 38.
26. Kenneth Silverman, *Edgar A. Poe: Mournful and Never-Ending Remembrance* (New York: Harper Perennial, 1991), 84.
27. Ibid., 83–84.
28. Poe submitted "Metzengerstein" to the *Saturday Courier* as late as December 1, 1831. Lee, "Absolute Poe," 775.
29. E. A. Poe, "Metzengerstein," in *The Unabridged Edgar Allan Poe* (Philadelphia: Running Press, 1983), 80.
30. Ibid., 84.
31. Poe, *Unabridged Edgar Allan Poe*, 85.
32. Ibid., 84.
33. Ibid., 85.
34. Edward Balleisen, *Navigating Failure: Bankruptcy and Commercial Society in Antebellum America* (Chapel Hill: University of North Carolina Press, 2001), 1–21; Bruce Laurie, *Working People of Philadelphia, 1800–1850* (Philadelphia: Temple University Press, 1980), 67–83, 119–24.
35. "Temperance," *Trumpet and Universalist Magazine* 5 (March 16, 1833), 150. Whether Poe drew the idea for his story from this quotation is unclear, but it was reprinted in Philadelphia's weekly newspaper, the *Saturday Courier*, on May 25, 1833. The

Saturday Courier was the same paper in which Poe published his first short story, "Metzengerstein," in 1832. Suggestively, in October 1833 Poe published the short story "MS. Found in a Bottle," which contains language very similar to "A Descent into the Maelström." "MS. Found in a Bottle" ends with the narrator being sucked into an enormous whirlpool in the South Seas.

36. Maria Lamas, *The Glass, or The Trials of Helen More: A Thrilling Temperance Tale* (Philadelphia: Martin E. Harmstead, 1849), 32.
37. Thomas Augst, "Temperance, Mass Culture, and the Romance of Experience," *American Literary History* 19 (2007): 297–323.
38. Poe, *Unabridged Edgar Allan Poe*, 687.
39. Ibid., 689.
40. Ibid., 690.
41. Ibid., 695.
42. Ibid., 694.
43. Ibid., 696.
44. Ibid., 698.
45. See chapter 3.
46. Augustine J. H. Duganne, *The Knights of the Seal, or The Mysteries of the Three Cities: A Romance of Men's Hearts and Habits*, 4th ed. (Philadelphia: G. B. Zieber, 1848).
47. Lamas, *Glass*.
48. Richard Bushman, *The Refinement of America: Persons, Houses, Cities* (New York: Vintage Books, 1993), 61–99, 280–312.
49. John F. Kasson, *Rudeness and Civility: Manners in Nineteenth-Century Urban America* (New York: Hill and Wang, 1990), 112–81; Karen Halttunen, *Confidence Men and Painted Women: A Study of Middle-Class Culture in America, 1830–1870* (New Haven, CT: Yale University Press, 1982), 96–97.
50. John Bowen Hamilton, "Robert Montgomery Bird, Physician and Novelist: A Case for Long-Overdue Recognition," *Bulletin of the History of Medicine* 44, no. 4 (1970): 315–42.
51. Charles Montgomery Bird, *Sheppard Lee: Written by Himself*, vol. 2 (New York: Harper and Brothers, 1836), 12–13.
52. Ibid., 21–22.
53. Ibid., 22.
54. Peter Simple, "The Confessions of a Drunkard," *Mechanic's Free Press*, February 21, 28, 1829.
55. *Quaker City: A Saturday Paper for Universal Circulation*, August 18, 1849. Cited by David Reynolds, *George Lippard, Prophet of Protest: Writings of an American Radical, 1822–1854* (New York: Peter Lang, 1986), 25.
56. David Reynolds, *Walt Whitman's America: A Cultural Biography* (New York: Vintage, 1996), 114.
57. Lippard, "Memoirs of a Preacher," 1.
58. George Lippard, *The Quaker City, or The Monks of Monk Hall: A Romance of Philadelphia Life, Mystery, and Crime*, ed. David S. Reynolds (Amherst: University of Massachusetts Press, 1995), 56.
59. Ibid., 54–55.
60. Ibid., 71.
61. Ibid., 93–112.
62. Bushman, *Refinement of America*, 282–87.

63. Lippard, *Quaker City*, 113–21.
64. Ibid., 122.
65. The success of the play at Barnum's museum became the prototype for the single long-running production that transformed standard practice in late nineteenth- and twentieth-century commercial theater. John W. Frick, *Theatre, Culture and Temperance Reform in Nineteenth-Century America* (New York: Cambridge University Press, 2003).
66. William Henry Smith, *The Drunkard, or The Fallen Saved! A Moral Domestic Drama: In Five Acts* (Boston: Jones's, 1847), 5–6. Although Smith produced the published script, the original author of the play was most likely the Unitarian minister and reformer John Pierpont.
67. Frick, *Theatre, Culture, and Temperance Reform*, 135.
68. Horticultural Hall, Zographicon, "Ten Nights in a Barroom" [1864?], Playbills collection, vol. 27, Library Company of Philadelphia.
69. "Grand Exhibition and Lecture . . . Dioramic Views," [1864–80?], Playbills collection, vol. 28, Library Company of Philadelphia.
70. Pliny Earle, "An Analysis of the Cases of Delirium Tremens Admitted into the Bloomingdale Asylum for the Insane from June 16th, 1821 to December 31st, 1844," *American Journal of the Medical Sciences* 15 (1848): 81–82.
71. "Animal Magnetism," *American Journal of the Medical Sciences* 21 (1837–38): 268–74.
72. *Minutes of the Meetings of the American Philosophical Society*, August 12, 1784; Charles d'Eslon, *Observations sur les deux rapports de MM. les commissaires nommés par Sa Majesté pour l'examen du magnétisme animal* (Philadelphia: Clousier, 1784).
73. Charles Willson Peale, *Charles Willson Peale: The Artist as Museum Keeper, 1791–1810*, ed. Lillian B. Miller, Sidney Hart, and Toby A. Appel, vol. 2, *The Selected Papers of Charles Willson Peale and His Family* (New Haven, CT: Yale University Press, 1988), 1022.
74. "Animal Magnetism," *Graham Journal of Health and Longevity* 3, no. 6 (1839): 94.
75. Daniel Drake, "Observations on Temulent Diseases," *American Medical Recorder* 2 (1819): 60–65; Daniel Drake, *A Discourse on Intemperance: Delivered at Cincinnati, March 1, 1828* (Cincinnati: Looker and Reynolds, 1828); Daniel Drake, *Analytical Report of a Series of Experiments in Mesmeric Somniloquism* (Louisville: F. W. Prescott, 1844).
76. *The Animal Magnetizer, or History, Phenomena and Curative Effects of Animal Magnetism: With Instructions for Conducting the Magnetic Operation by a Physician* (Philadelphia: James Kay Jr. and Brother, 1841).
77. Drake, *Analytical Report of a Series of Experiments in Mesmeric Somniloquism*, 49–50.
78. Ibid.
79. George Lippard, "Memoirs of a Preacher: A Revelation of Church and Home," *Quaker City News*, March 8, 1849, 1. Also, Lippard, *Quaker City*, 462.
80. William W. Pratt, *Ten Nights in a Bar-Room: A Drama in Five Acts, Dramatized from T. S. Arthur's Novel of the Same Name* (New York: Samuel French, 1860), 10.
81. Ibid., 28–29.
82. Pratt, *Ten Nights in a Bar-Room*, 22–23.
83. Horticultural Hall, Zographicon, "Ten Nights in a Barroom" [1864?], Playbills Collection, vol. 27, Library Company of Philadelphia.
84. *The New Family Book, or Ladies' Indispensable Companion and Housekeepers' Guide: Addressed to Sister, Mother, and Wife* (New York: Chambers, 1854), 26.

85. Pratt, *Ten Nights in a Bar-Room*, 36.
86. Ibid.

EPILOGUE

1. Ben Sharpsteen, *Dumbo*, RKO Radio Pictures, 1941, film.
2. Jack London, *John Barleycorn or, Alcoholic Memoirs* (London: Mills and Boon, 1913), 8–11.
3. Otto Messmer, *Felix Finds Out*, M. J. Winkler Company, 1924, film.
4. Sharpsteen, *Dumbo*.
5. Scott A. Sandage, *Born Losers: A History of Failure in America* (Cambridge, MA: Harvard University Press, 2005), 237.
6. Sarah Tracy, *Alcoholism in America: From Reconstruction to Prohibition* (Baltimore: Johns Hopkins University Press, 2005).
7. See the multiple subject headings under "Delirium Tremens" and "Alcoholism" in US Army, *Index-Catalogue of the Library of the Surgeon-General's Office*, 1st ser. (Washington, DC: Government Printing Office, 1880–95).
8. Jack S. Blocker, "Did Prohibition Really Work? Alcohol Prohibition as a Public Health Innovation," *American Journal of Public Health* 96, no. 2 (2006): 233–43.
9. Ron Roizen, "How Does the Nation's 'Alcohol Problem' Change from Era to Era? Stalking the Social Logic of Problem-Definition Transformations since Repeal," in *Altering American Consciousness: The History of Alcohol and Drug Use in the United States, 1800–2000*, ed. Sarah Tracy and Caroline Jean Acker (Amherst: University of Massachusetts Press, 2004).
10. Ibid.; Mariana Valverde, *Diseases of the Will: Alcohol and the Dilemmas of Freedom* (New York: Cambridge University Press, 1998), 96–119.
11. Michelle McClellan, "'Lady Tipplers': Gendering the Modern Alcoholism Paradigm, 1933–1960," in *Altering American Consciousness: The History of Alcohol and Drug Use in the United States, 1800–2000*, ed. Sarah Tracy and Caroline Jean Acker (Amherst: University of Massachusetts Press, 2004).
12. Trish Travis, *The Language of the Heart: A Cultural History of the Recovery Movement from Alcoholics Anonymous to Oprah Winfrey* (Chapel Hill: University of North Carolina Press, 2009), 21–59, Valverde, *Diseases of the Will*, 120–28.
13. Robin Room, Thomas Babor, and Jürgen Rehm, "Alcohol and Public Health," *Lancet* 365 (2005): 519–30.
14. Harris Isbell, H. F. Fraser, Abraham Wikler, R. E. Belleville, and Anna J. Eisenman, "An Experimental Study of the Etiology of 'Rum Fits' and Delirium Tremens," *Quarterly Journal of Studies on Alcohol* 16, no. 1 (1955): 1-33.
15. McClellan, "'Lady Tipplers.'"
16. William W. Pratt, *Ten Nights in a Bar-Room: A Drama in Five Acts, Dramatized from T. S. Arthur's Novel of the Same Name* (New York: Samuel French, 1860); Blake Edwards, *Days of Wine and Roses*, Warner Bros. Pictures, 1962, film.
17. Stanley Kubrick, *The Shining*, Warner Bros. Pictures, 1980, film.
18. T. J. Jackson Lears, *No Place of Grace: Antimodernism and the Transformation of American Culture, 1880–1920* (Chicago: University of Chicago Press, 1983), 300–307.
19. Sharpsteen, *Dumbo*.
20. Erika Dyck, *Psychedelic Psychiatry: LSD from Clinic to Campus* (Baltimore: Johns Hopkins University Press, 2008), 55.
21. Ibid., 53–78.
22. Jack Kerouac, *Big Sur* (New York: Penguin, 2011), 5–6.

23. Bill W[ilson], *Alcoholics Anonymous: The Story of How Many Thousands of Men and Women Have Recovered from Alcoholism*, 4th ed. (New York: Alcoholics Anonymous World Services, 2001), 13.
24. Ibid., 14.
25. Francis Hartigan, *Bill W.: A Biography of Alcoholics Anonymous Cofounder Bill Wilson* (New York: St. Martin's Griffin, 2001), 177–79.
26. Travis, *Language of the Heart*, 61–92.

INDEX